Recent Advances in Smart Design and Manufacturing Technology

Recent Advances in Smart Design and Manufacturing Technology

Guest Editors

Alessandro Greco
Donato Perfetto
Mario Caterino

MDPI

Basel • Beijing • Wuhan • Barcelona • Belgrade • Novi Sad • Cluj • Manchester

Guest Editors
Alessandro Greco
Department of Industrial Engineering
University of Salerno
Fisciano
Italy

Donato Perfetto
Department of Engineering
University of Campania "Luigi Vanvitelli"
Aversa
Italy

Mario Caterino
Department of Engineering
University of Campania "Luigi Vanvitelli"
Aversa
Italy

Editorial Office
MDPI AG
Grosspeteranlage 5
4052 Basel, Switzerland

This is a reprint of the Special Issue, published open access by the journal *Machines* (ISSN 2075-1702), freely accessible at: https://www.mdpi.com/journal/machines/special_issues/Design_and_Manufacturing.

For citation purposes, cite each article independently as indicated on the article page online and as indicated below:

Lastname, A.A.; Lastname, B.B. Article Title. *Journal Name* **Year**, *Volume Number*, Page Range.

ISBN 978-3-7258-6680-9 (Hbk)
ISBN 978-3-7258-6681-6 (PDF)
https://doi.org/10.3390/books978-3-7258-6681-6

Contents

About the Editors

Alessandro Greco

Alessandro Greco, PhD, is an Associate Professor at the University of Salerno. His current research focuses on mechanical design, experimental mechanics, and design for additive manufacturing, with contributions to national and European projects in the aerospace, automotive, robotics, digital twins, and advanced manufacturing sectors. He is the author of over 60 international publications and co-founder of a university spin-off company, and maintains active collaborations with both local industries and multinationals. He currently teaches machine design courses at the undergraduate, postgraduate, and doctoral levels.

Donato Perfetto

Donato Perfetto, PhD, is an Assistant Professor at the Department of Engineering, University of Campania Luigi Vanvitelli. His research focuses on mechanical design, structural modeling, composite materials, crashworthiness, impact behavior, and structural health monitoring (SHM). He has coordinated and contributed to several national and European research projects, including activities related to digital twins, additive manufacturing, and intelligent monitoring systems. He is the author of more than 56 scientific publications and actively collaborates with industrial partners on advanced simulation, material characterization, and smart sensing technologies.

Mario Caterino

Mario Caterino, PhD, is an Assistant Professor at the University of Campania Luigi Vanvitelli. After graduating with a degree in Mechanical Engineering, he received his PhD in Industrial and Information Engineering from the University of Campania Luigi Vanvitelli in 2021. He is currently a lecturer of several courses for undergraduate and postgraduate students at the Department of Engineering. He is the author of more than 60 international publications in Operations Management and Industrial Systems Engineering. He is a member of the Editorial Board of *The International Journal on Interactive Design and Manufacturing*.

Preface

This Reprint brings together eight peer-reviewed contributions originally published in the *Machines* Special Issue "Recent Advances in Smart Design and Manufacturing Technology". The collection reflects the rapid evolution of modern manufacturing toward digitally enhanced, human-aware, and sustainability-oriented production systems. Its scope spans from innovative approaches in additive and hybrid manufacturing to precision machining, human–machine interaction, and IoT-driven system optimization.

The aim of this work is to provide researchers, engineers, and practitioners with a concise overview of current advancements at the intersection of product design, materials processing, ergonomics, and smart factory paradigms. The contributions demonstrate how emerging technologies, such as 4D printing, advanced process monitoring, and intelligent scheduling, are reshaping industrial workflows, enabling more flexible and efficient manufacturing environments.

This Reprint is addressed to academics, industrial R&D professionals, and graduate students seeking insights into both novel manufacturing techniques and the critical role of human and digital integration in production. The authors involved represent diverse scientific communities and industrial perspectives, reinforcing the multidisciplinary nature of today's manufacturing research.

We would like to acknowledge the valuable work of the authors who trusted this Special Issue for disseminating their findings, the reviewers for their constructive evaluations, and the editorial team of *Machines* for their continuous support throughout the publication process.

As Guest Editors, we hope that this collection will inspire further research, collaboration, and innovation in the ongoing transformation of design and manufacturing technologies.

Alessandro Greco, Donato Perfetto, and Mario Caterino
Guest Editors

machines

MDPI

Article

A 4D-Printed Self-Folding Spatial Mechanism with Pre-Stressed Response Properties

Wencai Zhang [1] and Duanling Li [1,2,*]

[1] College of Mechanical and Electrical Engineering, Shaanxi University of Science and Technology, Xi'an 710021, China
[2] School of Automation, Beijing University of Posts and Telecommunications, Beijing 100876, China
* Correspondence: liduanlini@163.com

Abstract: Exploring the transformation of spatial mechanisms from their unfolded to controlled folding states to meet the requirements of various application scenarios has long been a hot topic in mechanical structure research. Although conventional spatial mechanisms can be designed to meet almost any application scenario, the design's complex and excessive combinations of structural components, kinematic pairs, and drive units are unavoidable. It introduces many problems, such as poor reliability, drive complexity, and control difficulties. Based on 4D printing technology, the design of self-folding spatial mechanisms that use pre-stressed response properties under predetermined thermal excitation to achieve different shrinkage ratios integrates the control and drive system and the structural components and kinematic pairs. It brings novel features of self-folding while effectively avoiding many problems associated with conventional mechanical design. Further, the pre-stressed response model introduces the self-folding spatial mechanisms' excitation, morphing, and driving investigation. Self-folding spatial mechanisms with different shrinkage ratios were prepared via fused deposition modeling, which verified the theoretical analysis and pre-stress response model and the design's correctness and feasibility by experiments. The existing 4D printing technology lacks a paradigmatic design method in the application field. Contrarily, this work organically combined the conventional mechanical structure design with materials and fabrication via fused deposition modeling. A systematic study of self-folding spatial mechanisms from structural design to morphing control was carried out. This design is expected to introduce a novel paradigm of 4D printing technology in conventional mechanical design and has considerable application prospects in spherical radar calibration mechanisms.

Keywords: 4D printing; self-folding; spatial mechanism; pre-stressed; scissor unit; fused deposition modeling

1. Introduction

Exploring the transformation of spatial mechanisms from their unfolded state to their controlled folded state to meet the needs of various application scenarios has long been a hot topic in mechanical structure research. Pinto's spatial mechanism design with a scissor unit marked the beginning of subsequent research on constructing a scissor unit-based spatial mechanism [1–4]. Hoberman improved the conventional scissor units and proposed angulated scissor units, which were constructed into the most classic Hoberman sphere and used in architecture [5–8]. The angulated scissor units demonstrate excellent spatial extensibility and shrinkage ratios in the construction of Hoberman spheres, in addition to the modularity of the scissor units, simplicity of construction, and low cost. For these reasons, researchers will maintain a keen interest in such units and mechanisms over the next few decades, leading to a wide range of applications in aerospace [9,10], transportation [11], and medicine [12]. Although conventional spatial mechanisms can be designed to meet almost any application scenario [13,14], the complex and numerous combinations of structural components, kinematic pairs, and drive units in the design are unavoidable, introducing a slew of problems, such as poor reliability, drive complexity, and control difficulties [15]. As a result, spatial mechanisms must keep their original powerful spatial extensibility while remaining structurally simple with reduced control, drive complexity, and increased reliability. This pressing matter must be addressed.

Four-dimensional printing has rapidly changed machinery engineering as we know it, allowing for more creativity and freedom in structure design, with numerous advantages over conventional manufacturing [16–18]. The material distribution and geometric parameters are changed by fused deposition modeling to impart pre-stressed response properties under predetermined thermal excitation and construct self-folding spatial mechanisms (SFSMs) with different shrinkage ratios [19–21]. Applying this new design, an SFSM integrates the control and drive system as well as the structural components and kinematic pairs into one and obtains self-folding of spatial volume or shape under predetermined thermal excitation. It brings a novel feature not available in conventional spatial mechanisms while effectively avoiding many problems associated with conventional mechanical design [22,23]. This work's main contributions are listed as follows:

1. An SFSM constructed of angulated scissor rod (ASR) and self-folding rod (SFR) was designed. The construction method and self-folding motion principle were analyzed, and the mathematical model of the SFSM in the self-folding motion was derived.
2. The pre-stressed response model of the SFSM was investigated, analyzing the thermodynamic properties of the material and identifying the response temperature. According to the design requirements combined with the material thermodynamic properties, the morphing patterns of the SFR were coupled. The influence of pre-stress on the folding morphing of the SFR under predetermined thermal excitation was explored to meet the requisite driving and folding.
3. Two SFSMs were printed and assembled, and the correctness and feasibility of the design, pre-stress response model, and theoretical analysis were verified by experiments.

2. Design of Self-Folding Spatial Mechanism

This section contains three subsections. First, the operational mode and composition of the SFSM are given. Second, the construction method and self-folding motion principle of the SFSM are analyzed. Finally, the mathematical model of the SFSM in the self-folding motion is derived.

2.1. Operational Mode and Composition

An SFSM with pre-stressed response properties under predetermined thermal excitation was designed, and its compositions and operating modes are shown in Figure 1. When no thermal excitation was applied, the SFSM showed an unfolded state. At this moment, the volume of space reached its maximum value. When the predetermined thermal excitation was applied, the SFSM gradually self-folded under pre-stress. Eventually, the volume of space reached its minimum value.

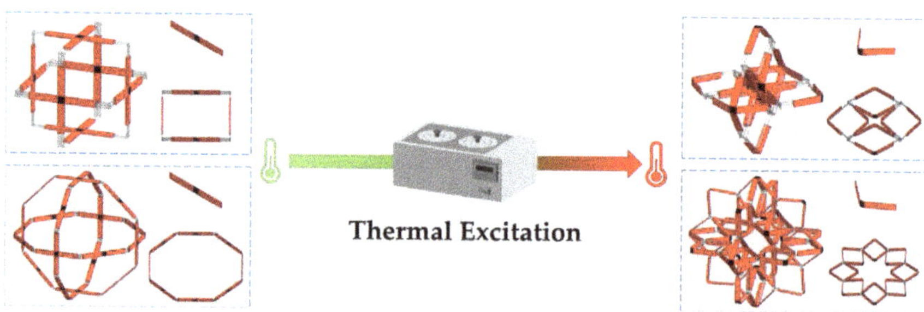

Figure 1. Schematic of the design and operating mode of the SFSM.

The pre-stressed response properties of the SFSM under predetermined thermal excitation were crucial to realizing the change in the volume of space. For this purpose, the SFR and ASR that constituted the SFSM were designed, as shown in Figure 2a. The ASR controlled the construction, shrinkage ratios, and morphing patterns of the SFSM based on a pre-designed structure and geometric arrangement. The SFR enabled the SFSM to obtain a self-folding feature based on the pre-stressed response properties under predetermined thermal excitation. When thermal excitation was applied, the SFR changed from an unfolded state to a folded state according to the pre-stressed response model and drove the ASR to fold the SFSM during this process.

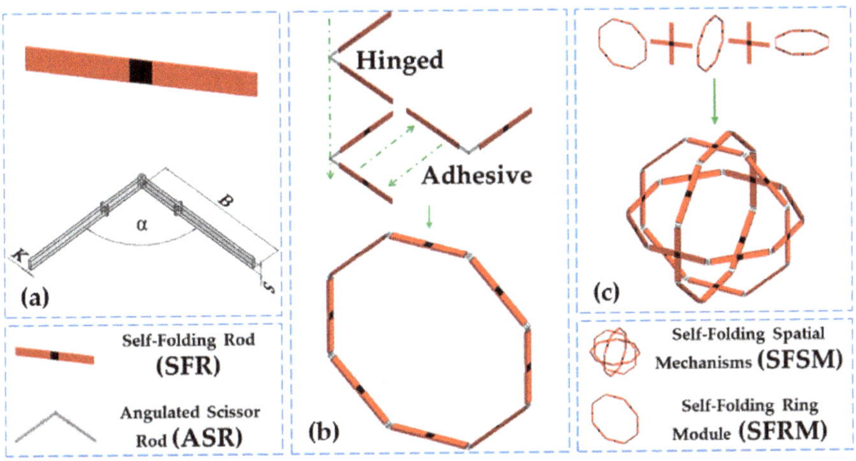

Figure 2. Schematic of the construction process of the SFSM: (**a**) the SFR and the ASR, (**b**) construct of the SFRM, (**c**) construct of the SFSM (left to right by order of appearance).

2.2. Construction Method and Self-Folding Motion Principle

As shown in Figure 3, the self-folding ring module (SFRM) was extracted from the constructed SFSM, and the ASR and SFR were extracted from the SFRM to establish a rectangular coordinate system. Point e was the hinge point, and point o was the intersection of a and d with the b and c lines. Two vertical lines from point e to ob and od were made, and points h and g were the feet of perpendiculars. Combined with Figure 3, we can prove that:

$$\because \angle aeb = \angle ced = \alpha$$
$$\because \angle cea \text{ is the common angle}$$
$$\therefore \angle ceb = \angle aeb - \angle aec = \angle ced - \angle cea = \angle aed$$
$$\because ce = ed = ae = eb$$
$$\therefore \triangle ceb \text{ and } \triangle aed \text{ are congruent isosceles triangles}$$
$$\because \text{The lines } eh \text{ and } eg \text{ are the perpendicular bisector of } \triangle ceb \text{ and } \triangle aed$$
$$\therefore \triangle ceh \cong \triangle beh \cong \triangle aeg \cong \triangle deg$$
$$\because \angle ceh = \angle beh = \angle aeg = \angle deg$$
$$\therefore 2\angle ceh + \angle cea = 2\angle aeg + \angle cea = \alpha$$
$$\therefore \angle ceh + \angle cea + \angle aeg = \alpha$$
$$\because \angle ohe + \angle oge + \angle heg + \angle hog = 2\pi$$
$$\because \angle ohe = \angle oge = \pi/2$$
$$\therefore \angle hog = \angle coa = \pi - \alpha$$

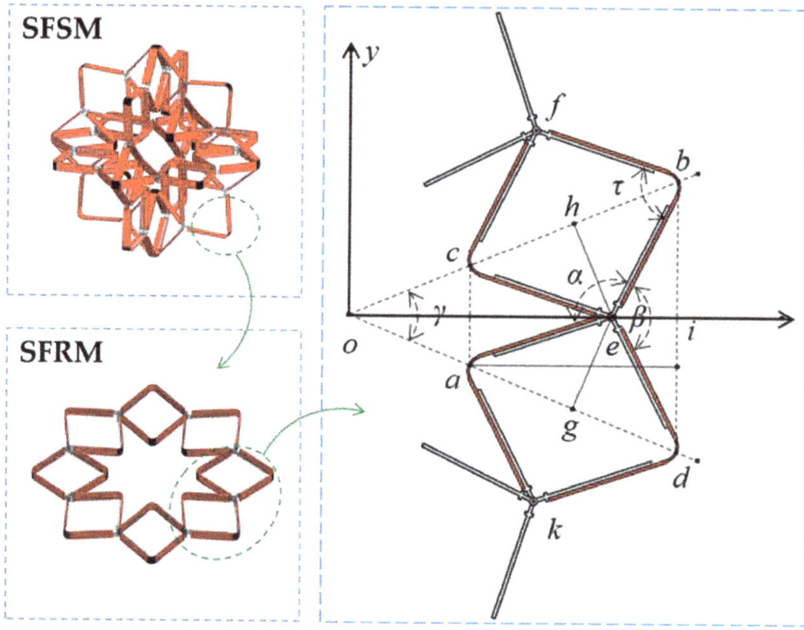

Figure 3. Schematic of the rectangular coordinate system of the ASRs and SFRs.

Let m be the number of rod groups (containing $2m$ SFRs and $2m$ ASRs) required to construct the SFRM.

$$\because \gamma = \angle hog = \angle coa = 2\pi/m$$

We know that:

$$\alpha = \pi - \frac{2\pi}{m} \wedge (m = 4k + 4 \wedge k \in Z) \tag{1}$$

Based on the above proof, it is known that the angle γ in a circular segment (such as angle dob) corresponding to each group of ASRs is always constant, and its value is only related to m. The value of m also determines the top angle α of the ASR. An SFRM can thus be constructed using m rod groups.

The construction method is as follows: the ASR is hinged, and the SFR adheres to the ASR. The serially connected SFRs and ASRs construct the SFRM in the two-dimensional plane, as shown in Figure 2b. As shown in Figure 2c, the paralleled adhered SFRMs construct the SFSM in three-dimensional space. Table 1 lists the SFRMs and SFSMs constructed at different values of m.

The previous paragraph detailed the SFSM's construction method. On this basis, the principle of self-folding motion was explained. In order to explain this principle, it is necessary to identify the motion features of the SFSM.

The SFRM motion features are first discussed. The ASR's relative rotation caused by the SFR's self-folding realizes the SFRM's self-folding. The SFR drives the ASR motion, and the hypothesis is that the motion is created by face contact between the ASRs extending at both ends. As a result, the SFR can be thought of as a planar revolving pair.

Table 1. Construction of SFRMs and SFSMs at different values of m.

m	Theoretical Shrinkage Ratio	Folded SFRM	Unfolded SFRM	Unfolded SFSM
4	1	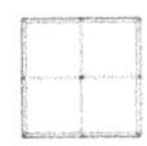	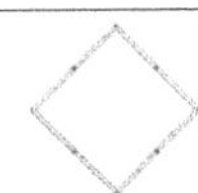	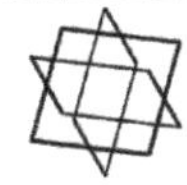
8	0.71	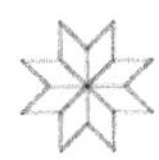	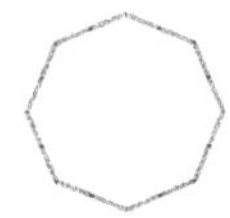	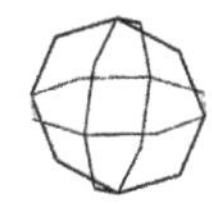
12	0.5		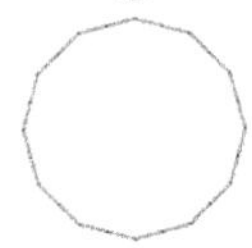	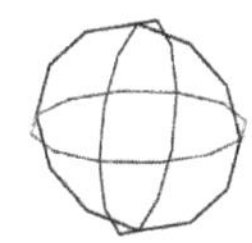
16	0.38	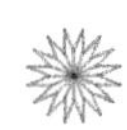	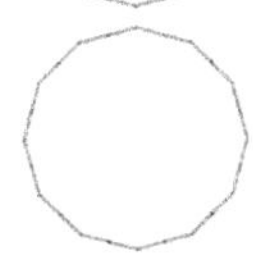	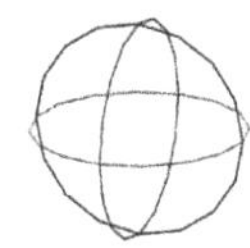

As shown in Figure 4, an SFRM was used with an m value of 12 to cut and expand into a plane kinetic chain along the line connecting points o and d.

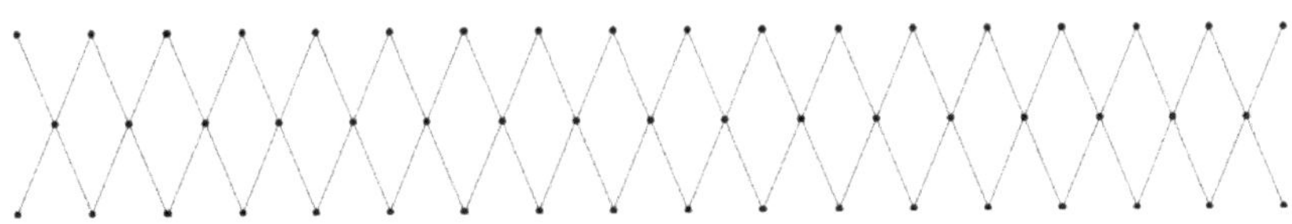

Figure 4. Plane kinetic chain with m value of 12.

Applying the loop connectivity matrix (LCM), we know that:

$$F_{LCM} = \begin{matrix} 1 & 1 & 0 & 0 & 0 \\ 0 & 1 & 1 & 0 & 0 \\ 0 & 0 & 1 & 1 & 0 \\ 0 & 0 & 0 & \ddots & 1 \\ 0 & 0 & 0 & 0 & m \end{matrix} \quad (2)$$

Let F be the degree of freedom. According to Equation (2), we concluded that the SFRM has a single degree of freedom.

The SFSM motion features are discussed next. The SFSM can be regarded as a three-dimensional structural combination of SFRMs perpendicular to each other, overlapping in the center of a circle. Li [24] analyzed similar mechanisms using screw theory and demonstrated that when the sub-kinetic chain has a single degree of freedom, the spatial mechanisms composed of the sub-kinetic chain have zero degrees of freedom in the fully expanded and contracted state and have a single degree of freedom during the motion. Thus, the SFSM also has similar degrees of freedom. As shown in Figure 5a, the SFR folds under predetermined thermal excitation, and the folding angle change process is $\tau_1 \rightarrow \tau_2 \rightarrow \tau_3$. SFR midpoints b and c move along the line ob, which drives the hinge point e between the ASR along the line eo movement. As shown in Figure 5b, the above process is presented as a contraction of serially connected SFRs and ASRs around the center of the SFRM's circle under predetermined thermal excitation. As shown in Figure 5c, the above process is presented as a contraction of mutually perpendicular SFRMs around the center of the SFSM's sphere under predetermined thermal excitation.

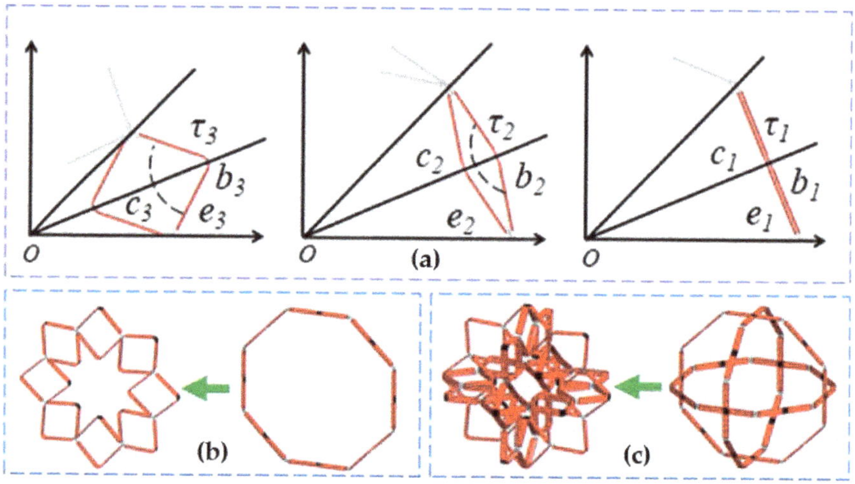

Figure 5. Schematic of self-folding: (**a**) SFR and ASR self-folding, (**b**) SFRM self-folding, (**c**) SFSM self-folding.

2.3. Mathematical Model

As shown in Figures 3 and 5, let the length of lines *ae*, *ce*, *be* and *de* be *B*. The angle *fbe* is defined as folding angle τ.

The mathematical model of the SFSM in the self-folding motion was derived:

The length of line *bd* is:

$$bd = 2B\sin(\frac{\beta}{2}) = 2B\sin(\frac{\gamma+\tau}{2}) = 2B\sin(\frac{\pi}{m}+\frac{\tau}{2}) \tag{3}$$

For the triangle *bod*, according to the law of cosines, we know that:

$$bd^2 = ob^2 + od^2 - 2ob\cdot od\cos\gamma \tag{4}$$

In the triangle *bod*, it is easy to know that *ob* is equal to *od*; bringing Equation (4), we know that:

$$bd^2 = 2ob^2(1-\cos\frac{2\pi}{m}) = 4ob^2\sin^2(\frac{\pi}{m}) \tag{5}$$

Letting Equation (3) equal Equation (5), we can solve the following:

$$ob(\tau) = \frac{B\sin(\frac{\pi}{m}+\frac{\tau}{2})}{\sin(\frac{\pi}{m})} \tag{6}$$

$$\tau = 2\mathrm{arcsin}(\frac{ob}{B}\sin\frac{\pi}{m}) - \frac{2\pi}{m} \tag{7}$$

When the folding angle reaches a minimum value, the SFSM is folded. Let R_{min} be the minimum circumscribed circle radius. Let τ_{min} be the minimum folding angle. In the triangle *bod*, it is easy to know that R_{min} is equal to *ob* and *od*; bringing Equation (7), we know that:

$$\tau_{\min} = 2\mathrm{arcsin}(\frac{R_{\min}}{B}\sin\frac{\pi}{m}) - \frac{2\pi}{m} \tag{8}$$

In the triangle *obe*, it is easy to know that:

$$\cos\angle boe = \frac{R_{\min}}{2}B^{-1} = \frac{\pi}{m} \tag{9}$$

Bringing Equation (9) into Equation (8), we know that:

$$\tau_{\min} = 2\arcsin(\sin\frac{2\pi}{m}) - \frac{2\pi}{m} = \frac{2\pi}{m}$$

The length of the line *oe* is:

$$oe = oi - ei = \frac{bd}{2}\cot(\frac{\gamma}{2}) - B\cos(\frac{\gamma}{2} + \frac{\tau}{2}) \tag{10}$$

Bringing Equation (4) into Equation (10), we can deduce:

$$oe(\tau) = B\cot(\frac{\pi}{m})\sin(\frac{\pi}{m} + \frac{\tau}{2}) - B\cos(\frac{\pi}{m} + \frac{\tau}{2}) \tag{11}$$

$$\tau = \arcsin(oe\sin\frac{\pi}{m}B^{-1}) \tag{12}$$

When the folding angle reaches a maximum value, the SFSM is unfolded. Let R_{max} be the maximum circumscribed circle radius. Let τ_{max} be the maximum folding angle. In the triangle *obe*, it is easy to know that R_{max} is equal to *oe*; bringing Equation (12), we know that:

$$\tau = \arcsin(R_{\max}\sin\frac{\pi}{m}B^{-1}) \tag{13}$$

In the triangle *obe*, it is easy to know that:

$$\sin\angle boe = \frac{B}{R_{\max}} = \frac{\pi}{m} \tag{14}$$

Bringing Equation (14) into Equation (13), we know that:

$$\tau_{\max} = 2\arcsin(1) = 2\pi$$

Its structural shrinkage ratio is:

$$\chi = \frac{R_{\min}}{R_{\max}} = \frac{ob(\tau_{\min})}{oe(\tau_{\max})} = \sin\left(\frac{2\pi}{m}\right) \tag{15}$$

From the above proof and calculations, we concluded that the m value not only controls the construction and shrinkage ratio of the SFSM but also controls the morphing patterns of the SFSM based on the folding angle τ. As a result, the ASR controls the construction, morphing, and shrinkage ratio of the SFSM by geometric arrangement based on a pre-designed number of m rod groups.

Conventional mechanical design methods considering only a single geometric feature and topology are not feasible for SFSM because of the pre-stressed response properties of controlling structural transformations under predetermined thermal excitation. This type of structure's design is not to be considered only in terms of construction methods and the principle of self-folding motion. Further research on the pre-stress response model based on 4D-printed materials and fabrication is needed to enable SFSM to obtain self-folding features based on the pre-stress response properties under predetermined thermal excitation.

3. Pre-Stressed Response Model

In this section, we began researching the pre-stressed response model of the SFSM. The SFR was a component with an integrated motion actuator and driver, which enabled the SFSM to obtain a self-folding feature based on the pre-stressed response properties under predetermined thermal excitation.

This section contains three subsections on the pre-stressed response model. First, the thermodynamic properties of the 4D-printed materials used to make the SFR were characterized. Second, the morphing patterns of the SFR were coupled by the combination and design of multi-material prefabricated components according to the self-folding design requirements of the SFSM and the characterization of material thermodynamic properties. Finally, in order to achieve the required driving and folding of the spatial mechanisms as much as possible, research on the folding morphing influences based on the pre-stress was carried out.

3.1. Characterization of Material Properties

The thermodynamic properties of different materials might be utilized to print and control the SFR. In addition, this work used predetermined thermal excitation as a means of self-folding activation. As a result, material property tests were conducted to characterize the material's thermodynamic properties and provide a relevant basis for subsequent research.

We selected four commercial elastomer materials based on thermoplastic polyurethane (TPU) (Dake, China) and one polymer material, polylactic acid (PLA) (Raise Premium, China). The dynamic thermodynamic properties of these five materials were analyzed using a dynamic thermo-mechanical analyzer (DMA-Q800, United States) in selected tensile mode. The practical test length of the PLA and TPU printed filaments was 10 mm, and the diameter was 1.75 mm. The test loading temperature range was 25 °C to 90 °C. The accuracy of the temperature loading was ±0.2 °C. The temperature rise rate was controlled at 2 °C/min during the test. The dynamic axial stretching rate was 1 Hz. The dynamic thermo-mechanical analyzer (DMA) test results included the changes in the storage modulus G and dielectric loss angle $Tan\delta$ with temperature T, as shown in Figure 6. The T_i, T_g, and T_h of PLA were 61.96, 68.02, and 73.57 °C, respectively. The G values for PLA corresponding to the three temperatures were 2458.76, 1375.28, and 637.75 MPa. The subscripts i, g, and h represented the beginning, transition, and end of PLA's glass transition phase. Similarly, the DMA test results for TPU showed that the T_g of TPU was below room temperature, and G values of TPU decreased slowly with the increasing temperature.

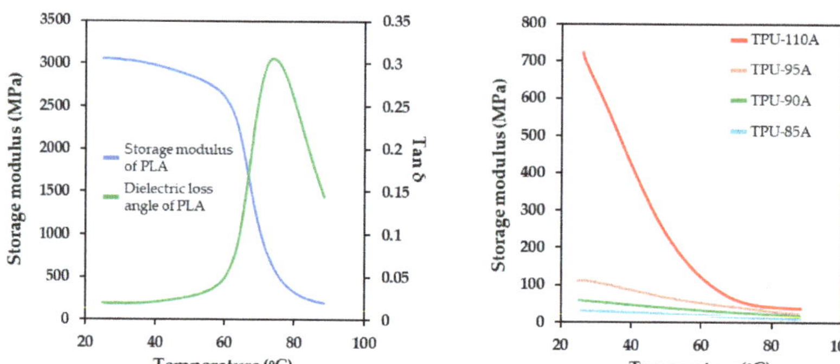

Figure 6. DMA test results: Storage modulus and dielectric loss angle of PLA (**left**) and storage modulus of TPU (**right**).

3.2. Coupling of Morphing Patterns

The SFR drove the SFSM to self-fold by bending patterns based on the pre-stressed response properties under predetermined thermal excitation. Therefore, this section discusses how to couple the SFR to produce bending by combining and designing multi-material prefabricated components according to the design requirements of the SFSM and the characterization of material thermodynamic properties.

As shown in Figure 7, the SFR was designed and printed utilizing TPU and PLA material by fused deposition modeling. This structure consisted of six layers, four of which were continuous and two that were split. The separation layers were designed to control the deformed part's width and compensate for the edge bending generated by the PLA layer.

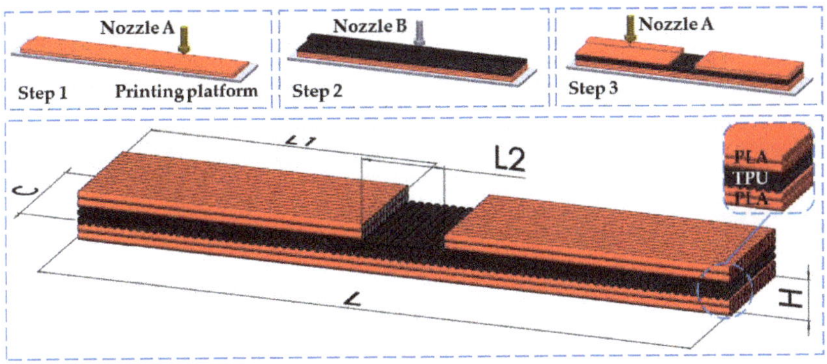

Figure 7. Schematic of the manufacturing process of the SFR.

First, how the SFR obtained the pre-stress response properties under predetermined thermal excitation was explained. Heating and squeezing the PLA filament during the printing process caused the polymer chains to stretch and align in the direction of that path and subsequently generate stress. They were stored in the printed material due to the constraining effect of the printing platform or previous layer. They were fixed layer by layer as the printing process cooled. When each PLA layer was removed from the printer and reheated above its glass transition temperature T_g, the pre-stress stored in the PLA layer was released and shortened along the printing direction while expanding slightly along the other two directions.

Second, it was explained how to couple the SFR to produce bending by combining and designing multi-material prefabricated components in accordance with the SFSM design requirements and the characterization of material thermodynamic properties. PLA layers with unidirectional filling patterns exhibit anisotropic morphing behavior, resulting in more significant anisotropic morphing behavior than multidirectional filling patterns [25,26]. For this reason, all PLA layers in this work were always printed in the same orientation. However, only single-layer PLA structures were used, which could produce unpredictable flex-torsion. The DMA test found that the glass transition temperature of TPU was generally lower than room temperature. The TPU elastic modulus was relatively stable over the T_h temperature range from room temperature to PLA, and it was assumed that it could not contract; it could only bend and slightly elongate. These properties were used, combining PLA and TPU in layers to print the SFR, which coupled the unpredictable morphing of PLA into bending. Although the TPU only played a restricted role in the SFR, its filling patterns still influenced bending. In order to investigate the influence of TPU filling patterns on morphing, the separation layer of SFR was removed. TPU filling patterns had a more noticeable influence on morphing when the separation layer was removed. As shown in Figure 8, another of our experimental results revealed that when the filling patterns of the TPU layers were perpendicular to the PLA layers and there was no separation layer, the structure exhibited the best bending.

3.3. Morphing Influence Based on Pre-Stress

Following the SFR-coupled bending, the influence of pre-stress on the SFR's folding under predetermined thermal excitation was investigated in this section to fulfill the requisite driving and self-folding. The previous discussion showed that storing and restricting pre-stress in the SFR influences bending. Therefore, if the morphing influence of the SFR was to be obtained, it was necessary to research the influence of the pre-stressed restricting capability of the TPU and the pre-stress storage capability of the PLA on the change in its folding angle τ.

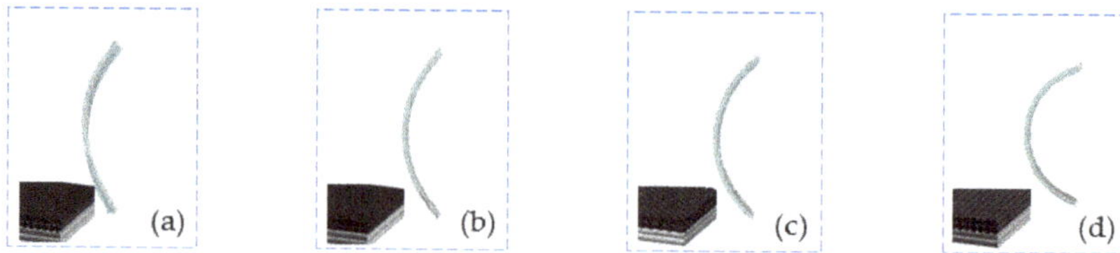

Figure 8. Experiments on the effect of different filling patterns of TPU layer on bending: (**a**) TPU material 90° cross alignment, (**b**) TPU 90° side-by-side alignment, (**c**) TPU material 45° cross alignment, (**d**) TPU 180° side-by-side alignment.

First, the pre-stressed restricting capability of the TPU is discussed. Four SFRs were printed and experimented with using four TPU materials. These SFRs were printed using a fused deposition modeling printer (Raise E2, Shanghai, China). Hot water was chosen as the activation medium for the experiments to ensure a uniform, accurate and fast heat application on the SFRs [27]. The glass transition temperature T_g of the PLA was selected from Figure 5, and the printing speed for all materials was set to 30 mm/s. The temperature setting of the water bath device (LICHEN-HH4, Shanghai, China) was kept constant. All SFRs used for the experiments were kept in water, and heating was stopped when they no longer exhibited visual signs of deformation. The printing parameters, structure size, and experimental parameters are shown in Table 2.

Table 2. Sample structure size and printing and experimental parameters of the SFR.

Structure Size/[mm]	H 1.2	C 10	L_2 10	L_1 45	L 100
Printing parameters		Layer height (mm)		0.2	
		Infill amount		100%	
		Extrusion width (mm)		0.4	
		Nozzle diameter (mm)		0.4	
		Printing platform temperature (°C)		30	
		PLA printing temperature (°C)		235	
		TPU printing temperature (°C)		200	
Experimental parameters		Activation medium		Water	
		Water bath temperature (°C)		68	
		Water bath time (s)		≥180	

Through experiments, it was found that TPU materials with higher storage modulus have highly pre-stressed restricting capability, causing weaker SFR drive and minor variation in folding angle τ, as shown in Figure 9. According to another experimental result in the literature [28,29], the lower the percentage of hard polymer segments supporting TPU materials, the more difficult it was to print them. After considering the printing quality and material properties, this work chose a single TPU-90A material to print the SFR.

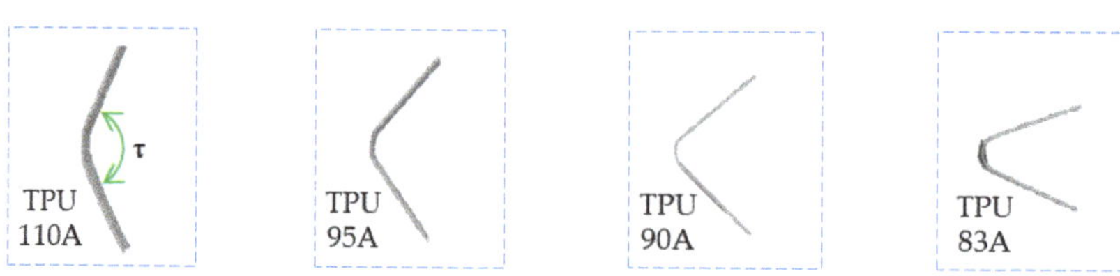

Figure 9. Experiment on the influence of TPU layer pre-stressed restricting capability on the change in folding angle τ.

Second, the pre-stress storage capability of the PLA is discussed. On the one hand, adjusting the printing speed caused different stretching of the PLA material during the extrusion process, resulting in different levels of pre-stress stored in the material. On the other hand, adjusting the printing platform temperature caused different mobility of the polymer chains in the PLA, resulting in different fixation times of their macroscopic shapes and stress relaxation effects. It also caused a difference in the levels of pre-stress stored.

Three experimental groups were established. The first group of 45 SFRs was printed with the printing platform temperature set to 30 °C and the PLA layer printing speed set to 150 mm/s. The second group of 45 SFRs was printed with the printing platform temperature set to 30 °C and the PLA layer printing speed set to 30 mm/s. The third group of 45 SFRs was printed with the printing platform temperature set to 70 °C and the PLA layer printing speed set to 150 mm/s. Only the above parameters were changed in the three experimental groups, and the other parameters were the same as in Table 1.

In each experimental group, five SFRs for each group were heated simultaneously in a water bath, and another group was heated for increasing intervals of 30 s. At the end of the time, the SFR was removed from the constant temperature water bath, cooled to room temperature, and placed on a scanning test bench to capture the surface shape. An optical 3D scanner (MetraSCAN 3D, Lévis, QC, Canada) was used to measure the folding angle τ after morphing. The collected data were combined to create corresponding 3D models for each experimental group to assess the experimental results more accurately and quantitatively. Figure 10 depicts the entire experimental process.

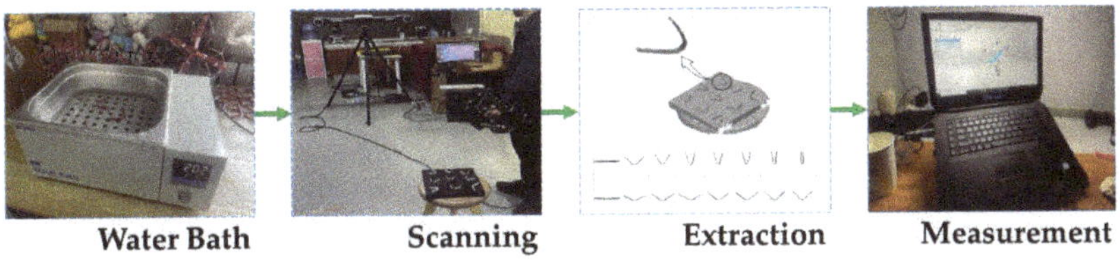

Figure 10. The 3D optical scanning processes.

The averages of the measured results of the folding angle τ are shown in Figure 11. The experimental results indicated the following:

1. Printing PLA materials at faster print speeds allowed for more significant stretching of the polymer chains during extrusion. Therefore, under the condition that the printing platform temperatures were constant, this approach allowed the SFR to maintain higher pre-stress, resulting in a broader range of folding angle variations and stronger drive capability.
2. Printing PLA materials on a lower temperature printing platform could quickly lock the polymer chains in a stretched state. Therefore, under the condition that the printing speed parameters were constant, this approach allowed the SFR to maintain higher pre-stress, resulting in a broader range of folding angle variations and stronger drive capability.
3. The folding angle of the SFR gradually decreased as the water bath time increased. Throughout the process, the self-folding was most evident within 2 min. After 3 min, the SFR samples no longer produced significant self-folding, and their folding angles remained stable. Therefore, the drive capability of the SFR continuously decreased with the increased water bath time.

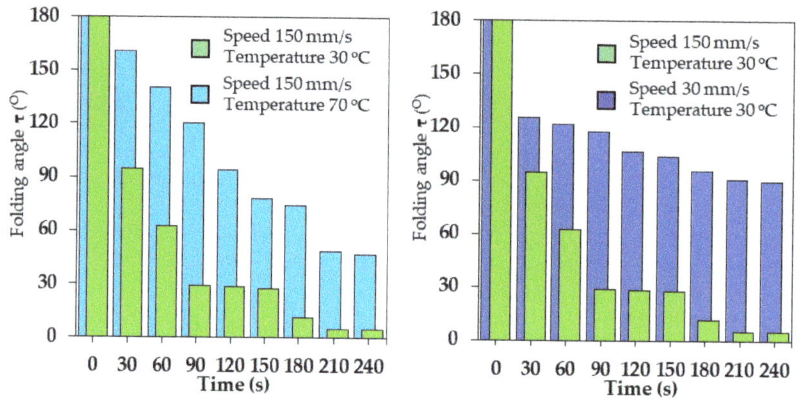

Figure 11. Experiment on the influence of PLA pre-stress storage capability on the change in folding angle τ.

4. Manufacture and Experiments

This section consists of two subsections. First, two SFSMs with m values of 4 and 8 were printed and assembled and named S4 and S8, respectively. In this class of mechanisms, S4 was chosen for its unique characteristics, while S8 was chosen for its universal characteristics. Second, the feasibility of self-folding of the SFSM was experimentally verified.

4.1. Printing and Assembly of the SFSM

The ASR and SFR were printed using a dual-nozzle fused deposition modeling printer (Raise E2, Chengdu, China) and assembled into SFSM. The S4 was constructed with 24 ASRs and 24 SFRs. The S8 was constructed with 48 ASRs and 48 SFRs.

The printing process for the ASR is described first because it controlled the constructions, shrinkage ratios, and morphing patterns of the SFSM based on a pre-designed structure and geometric arrangement. The ASR was printed using a common high-temperature resistant polycarbonate material (Raise, Premium PC). The ASR's length B was first determined. Equation (1) was then used to calculate the top angle α of the ASR based on the selected value of m. Equations (6) and (11) were used to calculate the theoretical maximum and minimum radius of the circumscribed circles based on the above two parameters. Because of the thickness S limitation, these mechanisms did not reach the theoretical shrinkage ratio. The specific printing parameters for the ASR in this work are shown in Table 3.

Table 3. Manufacturing parameters for the ASR.

Structure Parameters					
Number of groups m	Top angle α (°)		Side length B (mm)	Thickness K (mm)	Width S (mm)
4 8	90 135		60	3	5
Printing Parameters					
Printing platform temperature (°C)	Printing speed (mm/s)	Layer height (mm)	Infill amount	Extrusion width (mm)	Printing temperature (°C)
110	60	0.2	15%	0.4	235

The printing process for the SFR is described second because it enabled the SFSM to obtain a self-folding feature based on the pre-stressed response model. According to the experimental results, printing PLA materials on a lower temperature printing platform at faster print speeds allowed the polymer chains to stretch more significantly during extrusion and lock in the stretched state quickly, which helped to maintain high pre-stress to ensure that the SFR obtained significant driving and folding. The specific printing parameters of the SFR in this work are shown in Table 4.

Table 4. Manufacturing parameters for the SFR.

Structure Parameters								
Height H (mm)		Width C (mm)		Separation layer width L_1 (mm)		Separation layer spacing distance L_2 (mm)		Length L (mm)
1.2		10		45		10		100
Printing parameters								
Layer height (mm)	Infill amount	Extrusion width (mm)	Nozzle diameter (mm)	PLA		TPU		Platform temperature (°C)
				Speed (mm/s)	Temperature (°C)	Speed (mm/s)	Temperature (°C)	
0.2	100%	0.4	0.4	150	235	30	200	30

4.2. Experiments with the SFSM

The S4 and S8 were assembled. The experimental verification conditions and parameter settings were consistent with previous experiments. The experimental results are shown in Figure 12.

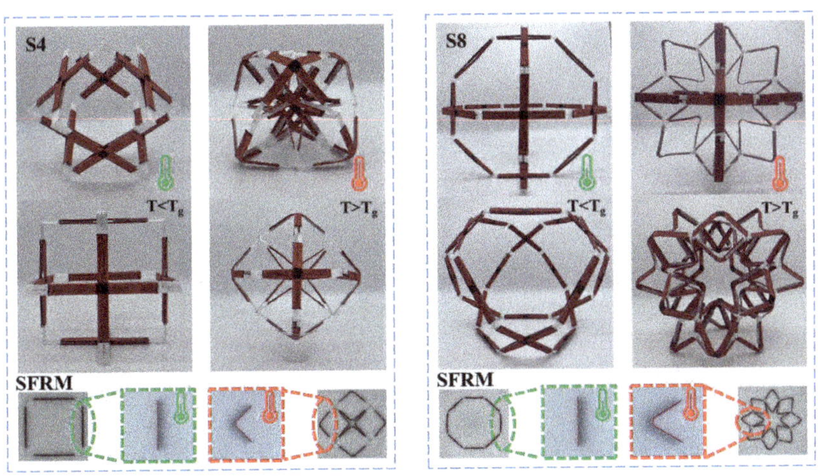

Figure 12. The self-folding experiment with the SFSM.

As shown in Figure 12a, the volume of S4 reached its maximum before thermal excitation ($T < T_g$), and the circumscribed circle radius R_{max} was about 100.39 mm. After thermal excitation ($T > T_g$), the SFR was self-folding, and the SFRM contracted. At this moment, the S4′s volume reached its minimum value, and the circumscribed circle radius R_{min} was about 96.47 mm. It should be noted that the actual and theoretical shrinkage ratio χ of this mechanism was 1 and about 0.96, respectively. The radius of the S4′s circumscribed circle changed negligibly during the self-folding process. However, after self-folding, its structure was significantly transformed. The square hexahedron made up of squares was transformed into an orthoctahedron made up of equilateral triangles.

As shown in Figure 12b, the volume of S8 reached its maximum before thermal excitation ($T < T_g$), and the circumscribed circle radius R_{max} was about 146.38 mm. After thermal excitation ($T > T_g$), the SFR was self-folding, and the SFRM contracted. At this moment, the S8′s volume reached its minimum value, and the circumscribed circle radius

R_{min} was about 129.13 mm. The actual and theoretical shrinkage ratio χ of this structure was about 0.71 and 0.88, respectively.

The experimental results demonstrated that the SFSM could realize the self-folding adjustment of the volume ratio under the predetermined thermal excitation, verifying the correctness and feasibility of the design and pre-stressed response model and theoretical analysis.

5. Conclusions

Based on 4D printing technology, SMSFs with different shrinkage ratios were prepared via fused deposition modeling, and the correctness and feasibility of the design, pre-stress response model, and theoretical analysis were verified by experiments.

Compared with the existing 4D printing technology, which lacks a paradigmatic design method in the application field, this work organically combined the conventional mechanical structure design with materials and fabrication via fused deposition modeling. A design was proposed based on the mutual integration of structural design and morphing control. It is expected that this design will introduce a novel paradigm of 4D printing technology into conventional mechanical design and have considerable application prospects in the design of spherical radar calibration mechanisms.

The material's distribution and geometric parameters were changed by fused deposition modeling to impart its pre-stressed response properties under predetermined thermal excitation and construct SFSMs with different shrinkage ratios. On the one hand, the structural components and kinematic pairs are combined into one, which effectively solves the problems of complex structure and excessive kinematic pairs in conventional spatial mechanical design. On the other hand, the drive and control system are combined into one, and the pre-stress response property under thermal excitation brought by 4D printing is used to replace the complex drive control system in the conventional spatial mechanical design while also bringing the novel feature of self-folding. This is not found in the conventional spatial mechanism.

In future work, more external thermal excitation sources (such as electric heating, optical heating, and magnetic field heating) will be introduced into the SFSM for regulation and experiment. Additionally, the hinges in the existing structures will be gradually replaced by smart materials to eventually realize SFSMs without mechanical connections.

Author Contributions: Conceptualization, W.Z. and D.L.; Data curation, W.Z.; Methodology, W.Z. and D.L.; Formal analysis, W.Z. and D.L.; Resources, D.L.; Investigation, W.Z. and D.L.; Software, W.Z.; Project administration, D.L.; Supervision, D.L.; Validation, D.L. and W.Z.; Writing—original draft preparation, W.Z.; Writing—review and editing, W.Z. and D.L. All authors have read and agreed to the published version of the manuscript.

Funding: This research was funded by the National Natural Science Foundation of China (Grant No. 52175019), and Beijing Natural Science Foundation of China (Grant No. 3212009).

Data Availability Statement: Not applicable.

Conflicts of Interest: The authors declare that there are no conflict of interest.

References

1. Escrig, F.; Valcarcel, J.P.; Sanchez, J. Deployable cover on a swimming pool in Seville. *J. Int. Assoc. Shell Spat. Struct.* **1996**, *37*, 39–79.
2. Freire-Tellado, M.J.; Muñoz-Vidal, M.; Pérez-Valcárcel, J. Bias deployable grids with horizontal compound scissor-like elements: A geometric study of the folding/deployment process. *Int. J. Space Struct.* **2021**, *37*, 22–36. [CrossRef]
3. Dai, J.S.; Li, D.; Zhang, Q.; Jin, G. Mobility analysis of a complex structured ball based on mechanism decomposition and equivalent screw system analysis. *Mech. Mach. Theory* **2004**, *39*, 445–458. [CrossRef]
4. Zhao, N.; Luo, Y.; Shen, Y. Design, modeling and validation of a deformable capsule-like crawling robot based on scissor elements. *SSRN Electron. J.* **2022**. *submitted*. [CrossRef]
5. Li, R.; Chen, H.; Choi, J.H. Topological assembly of a deployable Hoberman flight ring from DNA. *Small* **2020**, *17*, 124502–124511. [CrossRef]

6. Li, R.; Sun, X.; Chen, Y.; Yao, Y.; Ding, X. Design and analysis of reconfigurable deployable polyhedral mechanisms with straight elements. *J. Mech. Robot.* **2019**, *11*, 44502–44510. [CrossRef]
7. Hoberman, C. Tranformable: Building Structures that Change Themselves. In *Building Dynamics: Exploring Architecture of Change*, 1st ed.; Branko, K., Vera, P., Eds.; Routledge: London, UK, 2015; Volume 3, pp. 101–126.
8. Cai, J.; Xu, Y.; Feng, J. Kinematic analysis of Hoberman's linkages with the screw theory. *Mech. Mach. Theory* **2013**, *63*, 28–34. [CrossRef]
9. Lak, A.; Asefi, M. A new deployable pantographic lunar habitat. *Acta Astronaut.* **2022**, *192*, 351–367. [CrossRef]
10. Cao, W.; Xi, S.; Ding, H.; Chen, Z. Design and kinematics of a novel double-ring truss deployable antenna mechanism. *J. Mech. Des.* **2021**, *143*, 124502–124511. [CrossRef]
11. Zhao, N.; Luo, Y.; Wang, G.; Shen, Y. A deployable articulated mechanism enabled in-flight morphing aerial gripper. *Mech. Mach. Theory* **2022**, *167*, 104518. [CrossRef]
12. Dong, K.; Li, D.; Xue, X.; Xu, C.; Wang, H.; Gao, X. Workspace and accuracy analysis on a novel 6-UCU bone-attached parallel manipulator. *Chin. J. Mech. Eng.* **2022**, *35*, 2–13. [CrossRef]
13. Wang, R.; Dai, J. Design and analyses of a novel plane-space polyhedral reconfigurable metamorphic mechanism. *J. Mech. Eng.* **2013**, *49*, 29–35. [CrossRef]
14. Jia, P.; Li, D.; Zhang, Y.; Yang, C. A novel reconfigurable parallel mechanism constructed with spatial metamorphic four-link mechanism. *Proc. Inst. Mech. Eng. Part C J. Mech. Eng. Sci.* **2022**, *236*, 4120–4132. [CrossRef]
15. Kang, X.; Dai, J. Theoretical difficulties and research progresses of mechanism reconfiguration in mechanisms -evolution connotation, furcation principle, design synthesis and application of metamorphic mechanisms. *Chin. Mech. Eng. Soc.* **2020**, *31*, 57–71.
16. Coimbra, M.R.; Barbosa, T.P.; Vasques, C.M. A 3D-printed continuously variable transmission for an electric vehicle prototype. *Machines* **2022**, *10*, 84. [CrossRef]
17. Zhang, W.; Ge, Z.; Li, D. Evolution and emerging trends of 4D printing: A bibliometric analysis. *Manuf. Rev.* **2022**, *9*, 30. [CrossRef]
18. Khalid, M.Y.; Arif, Z.U.; Noroozi, R.; Zolfagharian, A.; Bodaghi, M. 4D printing of Shape Memory Polymer Composites: A review on fabrication techniques, applications, and future perspectives. *J. Manuf. Process.* **2022**, *81*, 759–797. [CrossRef]
19. Khalid, M.Y.; Arif, Z.U.; Ahmed, W.; Umer, R.; Zolfagharian, A.; Bodaghi, M. 4D printing: Technological developments in robotics applications. *Sens. Actuators A Phys.* **2022**, *343*, 113670. [CrossRef]
20. Arif, Z.U.; Khalid, M.Y.; Noroozi, R.; Sadeghianmaryan, A.; Jalalvand, M.; Hossain, M. Recent advances in 3D-printed polylactide and polycaprolactone-based biomaterials for tissue engineering applications. *Int. J. Biol. Macromol.* **2022**, *218*, 930–968. [CrossRef]
21. Khalid, M.Y.; Arif, Z.U.; Ahmed, W. 4D printing: Technological and manufacturing renaissance. *Macromol. Mater. Eng.* **2022**, *307*, 2200003. [CrossRef]
22. Arif, Z.U.; Khalid, M.Y.; Ahmed, W.; Arshad, H. A review on four-dimensional (4D) bioprinting in pursuit of advanced tissue engineering applications. *Bioprinting* **2022**, *27*, e00203. [CrossRef]
23. Arif, Z.U.; Khalid, M.Y.; Zolfagharian, A.; Bodaghi, M. 4D bioprinting of smart polymers for biomedical applications: Recent progress, challenges, and future perspectives. *React. Funct. Polym.* **2022**, *179*, 105374. [CrossRef]
24. Li, D.; Dai, J.; Zhang, Q.; Jin, G. Mobility of a kind of metamorphic mechanism-magic ball. *Chin. J. Mech. Eng.* **2002**, *38*, 12–16. [CrossRef]
25. van Manen, T.; Janbaz, S.; Zadpoor, A.A. Programming 2D/3D shape-shifting with hobbyist 3D printers. *Mater. Horiz.* **2017**, *4*, 1064–1069. [CrossRef] [PubMed]
26. Kačergis, L.; Mitkus, R.; Sinapius, M. Influence of fused deposition modeling process parameters on the transformation of 4D printed morphing structures. *Smart Mater. Struct.* **2019**, *28*, 105042. [CrossRef]
27. Hu, G.F.; Damanpack, A.R.; Bodaghi, M.; Liao, W.H. Increasing dimension of structures by 4D printing shape memory polymers via fused deposition modeling. *Smart Mater. Struct.* **2017**, *26*, 125023. [CrossRef]
28. Qi, H.J.; Boyce, M.C. Stress–strain behavior of thermoplastic polyurethanes. *Mech. Mater.* **2005**, *37*, 817–839. [CrossRef]
29. Kim, K.; Park, J.; Suh, J.-H.; Kim, M.; Jeong, Y.; Park, I. 3D printing of multiaxial force sensors using carbon nanotube (cnt)/thermoplastic polyurethane (TPU) filaments. *Sens. Actuators A Phys.* **2017**, *263*, 493–500. [CrossRef]

machines

MDPI

Article

Combined Use of sEMG and Inertial Sensing to Evaluate Biomechanical Overload in Manufacturing: An On-the-Field Experience †

Maria Grazia Lourdes Monaco [1], Lorenzo Fiori [2,3,‡], Agnese Marchesi [4,‡], Mariarosaria Muoio [5], Elpidio Maria Garzillo [6,*], Francesco Caputo [7], Nadia Miraglia [5], Monica Lamberti [5], Alessio Silvetti [8] and Francesco Draicchio [8]

1 Occupational Medicine Unit, University Hospital of Verona, 37134 Verona, Italy; mariagrazialourdes.monaco@aovr.veneto.it
2 Department of Physiology and Pharmacology, PhD Program in Behavioral Neuroscience, Sapienza University of Rome, Viale dell'Università, 30, 00185 Rome, Italy
3 Department of Occupational and Environmental Medicine, Epidemiology and Hygiene, Italian Workers' Compensation Authority (INAIL), Monte Porzio Catone, 00078 Rome, Italy
4 Microsoft Italia, Viale Pasubio, 21, 20154 Milano, Italy
5 Department of Experimental Medicine, University of Campania Luigi Vanvitelli, Via Santa Maria di Costantinopoli, 16, 80138 Naples, Italy
6 Department of Prevention, Abruzzo Local Health Authority, 67100 L'Aquila, Italy
7 Department of Engineering, University of Campania Luigi Vanvitelli, Via Roma, 29, 81031 Aversa, Italy
8 INAIL Research Center, Via di Fontana Candida, 1, Monte Porzio Catone, 00078 Rome, Italy
* Correspondence: egarzillo@asl1abruzzo.it; Tel.: +39-0864-499623

† This paper is an extended version of our paper published in Monaco, M.G.L.; Fiori, L.; Marchesi, A.; Greco, A., Ghibaudo, L.; Spada, S.; Caputo, F., Miraglia, N., Silvetti, A., Draicchio, F. Biomechanical Overload Evaluation in Manufacturing: A Novel Approach with sEMG and Inertial Motion Capture Integration. In: Bagnara, S., Tartaglia, R., Albolino, S., Alexander, T., Fujita, Y. 2019 (eds) Proceedings of the 20th Congress of the International Ergonomics Association (IEA 2018). IEA 2018. Advances in Intelligent Systems and Computing, vol 818. Springer, Cham. https://doi.org/10.1007/978-3-319-96098-2_88.

‡ These authors contributed equally to this work.

Abstract: Biomechanical overload is considered a significant occupational risk in manufacturing and a potential cause of musculoskeletal disorders. This research aims to introduce new methodologies for the quantitative risk evaluation of biomechanical risk by combining surface electromyography with a motion acquisition system based on inertial measurement units. Due to the lack of experimental data in the literature acquired in a real industrial environment during the working shift, an on-the-field study regarding an automotive assembly line workstation has been carried out in collaboration with Fiat Chrysler Automobiles Italy S.p.A. Data related to the trunk flexion forward and the erector spinae muscle activity have been acquired for several consecutive working cycles by considering three different workers. Data analyses indicated kinematic and muscular activity patterns consistent with those expected and that the proposed wearable technologies can be integrated and used simultaneously during work activities. Furthermore, the results demonstrated data repeatability, strengthening the feasibility and usefulness of the combined use of kinematic and electromyography technologies to assess biomechanical overload in production lines. This study could lay the bases for the future definition of a method for assessing biomechanical overload due to awkward postures.

Keywords: surface electromyography; inertial sensors; biomechanical overload; experimental data analyses

1. Introduction

Musculoskeletal disorders (MSDs) are the most frequent occupational disorders in the European Union: they affect workers in all sectors and occupations, and they are the most important causes of long-term sickness absences with effects and costs not only on workers themselves but also on the society as a whole [1]. Several reasons have been identified as follows, and even if the work-related biomechanical load is not the only causative factor, it is likely to constitute a significant part of it: posture, repetitive movements, heavy lifting, awkward postures, exposure to cold temperature and insufficient recovery time, as well as psychosocial risk factors [2,3]. All those factors must be considered to estimate exposure [4].

The prevention of MSDs, a cornerstone of ergonomics and a challenge in industrial settings, must be integrated with correctly evaluating and managing biomechanical overload. Two risk assessment approaches allow for analysing the amount of discomfort and postural stress: observational and instrument-based techniques [5,6].

Traditionally, working postures and movements have been assessed using various observational protocols and checklists, such as the Ovako Working Postures Assessment System (OWAS) [7,8], the Rapid Upper Limb Assessment (RULA) [9], and the Rapid Entire Body Assessment (REBA) [10]. In some occupational contexts, e.g., the automotive industry, specific working methods have been developed, such as the Ergonomic Assessment Worksheet (EAWS) [11]. These assessment tools use on-the-job observation or video recordings to classify the ranges within which each body segment falls, with obvious limitations in characterising physical exposure: subjectivity, observer bias, low accuracy, long analysis periods, and the need for highly trained observers. Their internal and external validity has also been questioned. Many observational tools for biomechanical risk assessment have been developed; these tools only require a little equipment, other than an evaluation sheet and pencil, and moderately agree with technical measurements [5,12]. These methods are also vulnerable to errors: the most significant discrepancies arise in estimating the applied forces and the posture, and their inter-rater reliability might vary a lot [13].

In the Industry 4.0 era, several research groups have been interested in the application of new technologies in the field of ergonomics, also through the combination of different methods to allow for the use of quantitative biomechanical measures, which are more precise and reliable, and to obtain detailed and accurate values for jobs with varying tasks of work [13–19]. In recent years, wearable sensors have been used for quantitative instrumental-based biomechanical risk assessments to prevent work-related musculoskeletal disorders (WMSDs) [20]. Surface electromyography (sEMG) is considered an important and helpful tool for the quantitative evaluation of biomechanical overload and offers the possibility of obtaining 'inline' information, highly relevant from several ergonomic points of view [21,22]; sEMG is a non-invasive method and, for this reason, it can be used during the execution of a work task [23]. Several methods evaluate the range of motion during professional activities, for instance real-time measurement could be conducted using sensors attached to the worker's body. For industrial applications, motion capture systems record workers' gestures to assess ergonomic risk and improve working conditions objectively. Motion capture systems consisting of Inertial Measurement Units (IMU) represent the best solution for ergonomic applications in a real occupational setting since they do not hinder working activities and they are not bulky like vision systems, even if data could be less accurate than those ones. Several researchers have introduced IMU devices to measure workers' body motion [14,17,24,25]. However, the equipment mentioned above suffers from possible electromagnetic interference, which occurs widely in industrial environments [17].

Acquiring the real working condition data may allow us to evaluate the dynamic postural aspects of the single worker's activities, which observational pencil and paper methodologies would not provide. This feature can allow us to identify the real contribution to each occupational task's biomechanical overload and assess the effectiveness of any preventive and corrective interventions through pre- and post-measurements. Moreover, a combined approach throughout assessing muscle activity and a kinematic evaluation could lead to a comprehensive assessment of the dynamic effects of the workers' postures/activities. In the literature, some approaches integrate multiple technologies (sEMG, IMUs, and videotaping) to objectify the different factors of biomechanical overload [15,16,18]. These approaches are innovative, and there needs to be evidence of analyses of data collected in the manufacturing environment during the normal production processes. An interesting experiment was conducted, in the laboratory, by Poitras et al. [26], who studied the validity of using wearable sensors at the shoulder joint, combining EMG and IMU sensors. Although they highlighted the suitability of the combined use of the sensors during a work task simulation, the authors emphasised the need to validate their use in the workplace, in real work situations. Merino et al. evaluated the shoulder biomechanical overload in three workers performing banana processing tasks using inertial sensor motion capture (Xsens) and EMGs [27]. The methods used in the evaluation provided useful data on the possible relationship between awkward posture and the occurrence of fatigue and musculoskeletal disorders. This study also supports the need to obtain on-field data.

This on-the-field study is within the framework of a 2019 PhD project focused on the role of sEMG in biomechanical overload assessment in the automotive manufacturing setting. Some data were only preliminarily presented in the 2019 IEA publication [28]. This study aims to illustrate a methodological approach for quantitatively assessing biomechanical overload risk based on the combined use of sEMG and an IMU-based wearable motion capture system for collecting experimental data during work activity in automotive manufacturing production lines.

2. Materials and Methods

2.1. Setting and Subjects

The study was conducted in collaboration with Fiat Chrysler Automobiles Italy S.p.A., at the assembly shop of the plant located in Melfi, according to a protocol previously described [26]. Three male workers were enrolled (mean age 36 ± 12 years (SD); mean mass 79 (±14) kg (SD); height 173.7 ± 5.1 cm; working seniority 14.7 ± 11.0 years (SD)). They reported no prior cases of low back pain or surgeries. The research was performed following the ethical standards laid down in the 1964 Declaration of Helsinki and its later amendments. Ethical approval is not necessary because the workers' measurements were performed within the mandatory risk assessment process and according to Italian laws concerning the protection of workers exposed to occupational risks (Italian Decree no.81/2008).

2.2. Working Activity Description

The following figures show a phase of the activity carried out by a worker on the right side (Figure 1) and on the left side (Figure 2) of the workstation, where the central cabinet is assembled inside the cabin using screws, dowels, and cables. The cycle duration is approximately one minute (58 s). Experimental data have been acquired for about forty consecutive working cycles per worker. The activity was studied by analysing the Standard Operating Procedure (SOP) and with the support of the videos recorded by three different cameras: two cameras are located behind and to the worker's side, and another one is located on the worker's goggles. This last camera is integrated and synchronised with the motion capture system.

The kinematic and electromyographic analyses focused only on the trunk in this study. The choice to investigate this region is due to the preliminary observational investigations of the working task that suggested the lumbar spine district was most overloaded.

2.3. Instrumentation, Procedure, and Data Acquisition

2.3.1. Acquisition and Processing of sEMG Signals

A six-probe electromyography device (FreeEMG, BTS SpA, Milan, Italy) was used to record the electrical activity of the muscles. Each of them has a 100 dB CMRR instrumentation amplifier, a Hamming band-pass filter, a sampling frequency of 1 kHz, an analog-to-digital conversion system, and a wireless data transfer system (Wi-Fi). According to the recommendations of the Atlas of Muscle Innervation Zones [29], the probes were placed over the muscles engaged in the research using pre-gelled Ag/AgCl electrodes (H124SG, Kendall ARBO, Donau, Germany). Specifically, electrical activity was collected bilaterally from the paravertebral muscles: Erector Spinae Thoracic (EST), Erector Spinae Lumbar (ESL), and Multifidus (M), of which the landmarks for electromyographic measurements were identified, positioning the electrodes following the indications of the innervation zone atlas as given in the literature [29] (Figure 3).

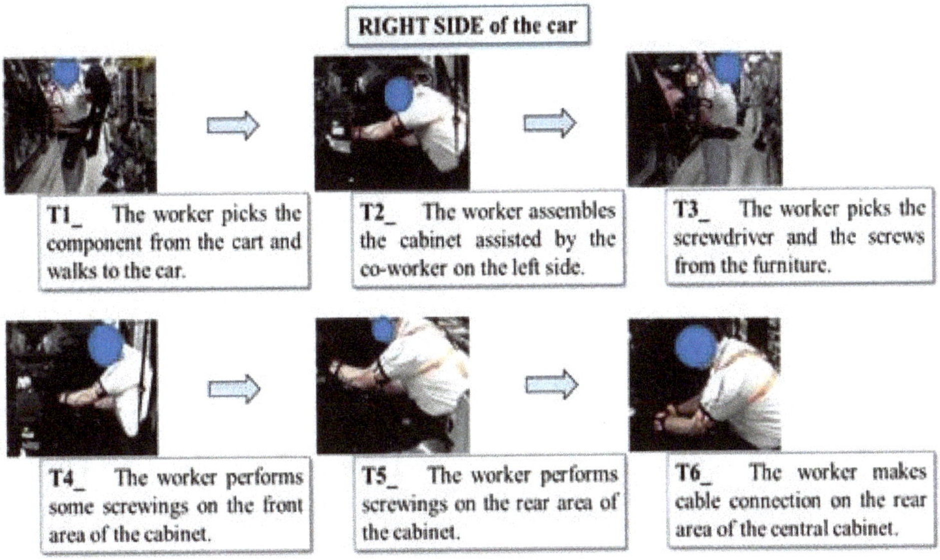

Figure 1. Working activity on the right side of the workstation.

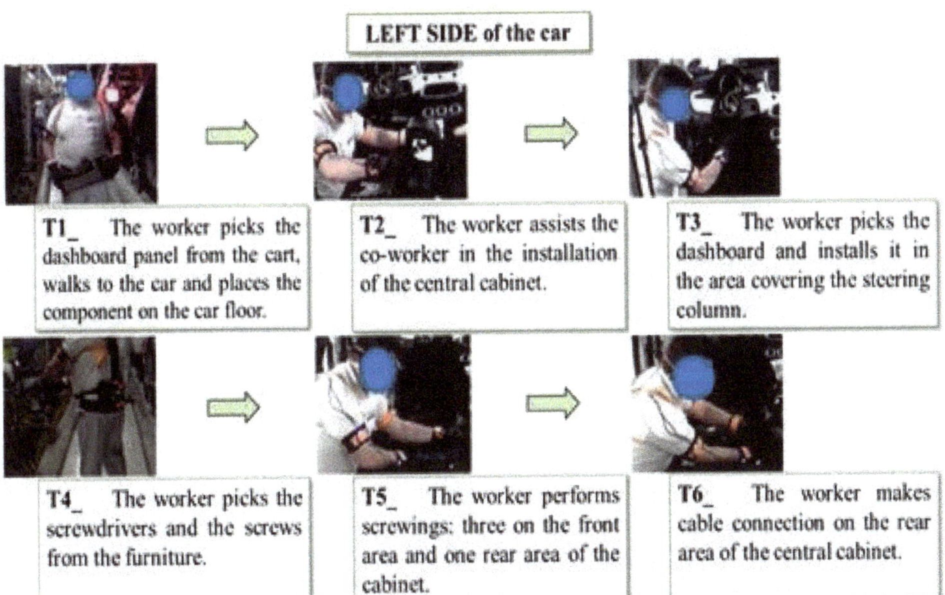

Figure 2. Working activity on the left side of the workstation.

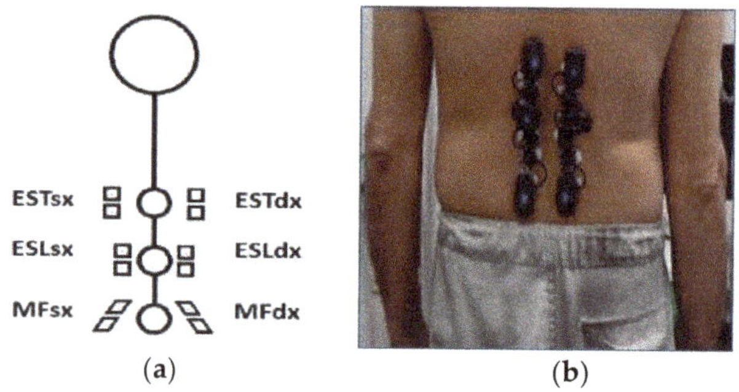

Figure 3. Example of the correct positioning of the electrode for EST assessment (**a**) and its placement on the operator's back (**b**).

The muscles' maximum voluntary contractions (MVCs), recorded before the work activity began, were used to calculate the peak amount of muscle activation that would serve as a benchmark during the signal processing stage. The patient extended his back for 15 s while lying on the abdomen with all his strength (Figure 4). After a three-minute break between each trial, each MVCs assignment was repeated three times, and the average value was calculated [21,30].

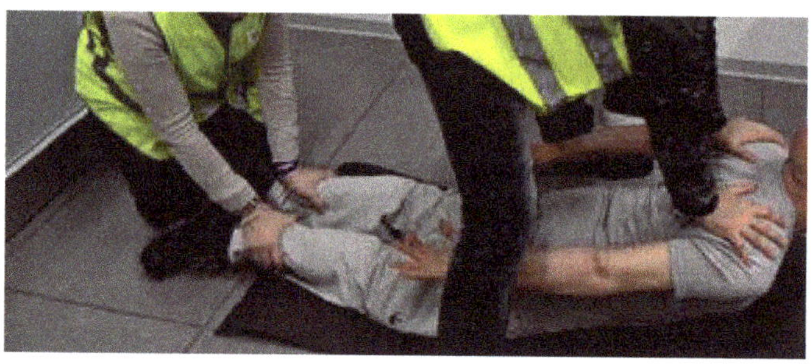

Figure 4. Isometric contraction test (maximum voluntary contraction) for the Erector Spinal muscles. The figure shows the prone position of the worker who performed a back extension involving the whole paravertebral musculature.

An algorithm created in MATLAB software (verses 9.3.0, MathWorks, Natick, MA, USA) was used to analyse the collected sEMG signals [29]. To lessen motion artefacts (electrode skin) and other high-frequency noise components, the electromyography signals were first filtered with an a%th-order Butterworth IIR digital pass filter in the study's target frequency range (30–450 Hz). The muscular activity profile was then extracted using an adaptive sEMG envelope extraction algorithm. [31]. Finally, the sEMG signal envelope was expressed as a percentage of the MVCs of each muscle and time-normalized from 0 to 100% of the work cycle. The root mean square (RMS) index was calculated to obtain an overall index of muscle activity:

$$RMS = \sqrt{\frac{1}{N}\sum_{i=1}^{N}|x_i|^2} \quad (1)$$

where N is the total number of samples and x_i is the i-th sample value.

2.3.2. Body Motion Study

An inertial motion capture wearable system has been used to study the body's mobility. The system was created at the Luigi Vanvitelli Engineering Department of the University of Campania and comprised several micro–Inertial Measurement Units (IMUs) [25]. The upper-body configuration in this study required the development of a system made up of two independent modules. Four IMUs make up each module, placed on the pelvis, boot, arm, and forearm, respectively (Figure 5). A Raspberry Pi that is powered by a battery records and pre-processes data.

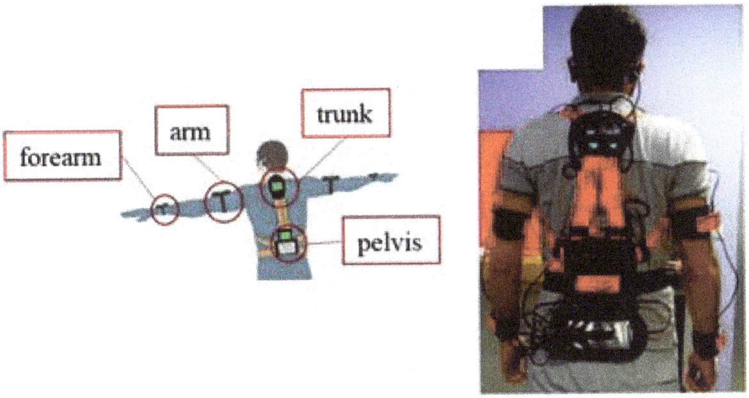

(a) (b)

Figure 5. Wearable motion capture system, in upper-body configuration: (**a**) scheme; (**b**) equipped by the worker.

The attitude estimation is based on a Kinematic Extended Kalman Filter [25] and provides both attitude data (in terms of quaternions and Euler angles), per each IMU, and posture angles trends over time: rigid pelvis rotation; flexion forward, lateral flexion and torsion of the trunk; elevation, lateral flexion and rotation of the arm; flexion and rotation of the forearm. The alignment of axes between IMUs and body segments is assured at the initial time by a software reset with the acquisition of the initial conditions from inertial sensors. The system's accuracy was tested and verified during experimental tests in a laboratory. Posture angles data were compared with those provided by the optical motion capture system SMART-DX by BTS Engineering®.

To make the estimation less subject to possible electromagnetic interferences, typical of industrial environments, the on-board Kalman filter was augmented with an adaptive virtual magnetometer reset when a significant difference in the magnitude of the magnetic vector was sensed. To match the postures data and the electromyographic data, only the flexion forward angle of the trunk has been considered in this study.

2.4. Results Analysis Methodology

The analysis of experimental data included the evaluation of the normality of the distributions and then applying comparison tests between averages/mediums. The Shapiro–Wilk test was used to evaluate the normality distribution of the data (due to a sample size of less than 30). In normal distributions, parametric comparison tests (*t*-test or ANOVA) were used; for non-normal distributions, non-parametric tests were used (Mann–Whitney for comparing two distributions, Kruskal–Wallis test for the comparison of more than two distributions). A *p*-value of <0.05 was considered statistically significant, and post hoc analyses were performed using a paired *t*-test with Bonferroni's corrections when significant differences were observed in the ANOVA. The statistical analysis was performed using MATLAB software (verses 9.3.0, MathWorks, Natick, MA, USA).

3. Results

3.1. Electromyographic Signal Analysis

The raw signals, recorded by each electromyographic electrode placed on the enrolled workers, were acquired and processed to obtain the sEMG envelope. Figure 6 shows EMG signals envelopes for the working task at the right and left side of the workstation, respectively.

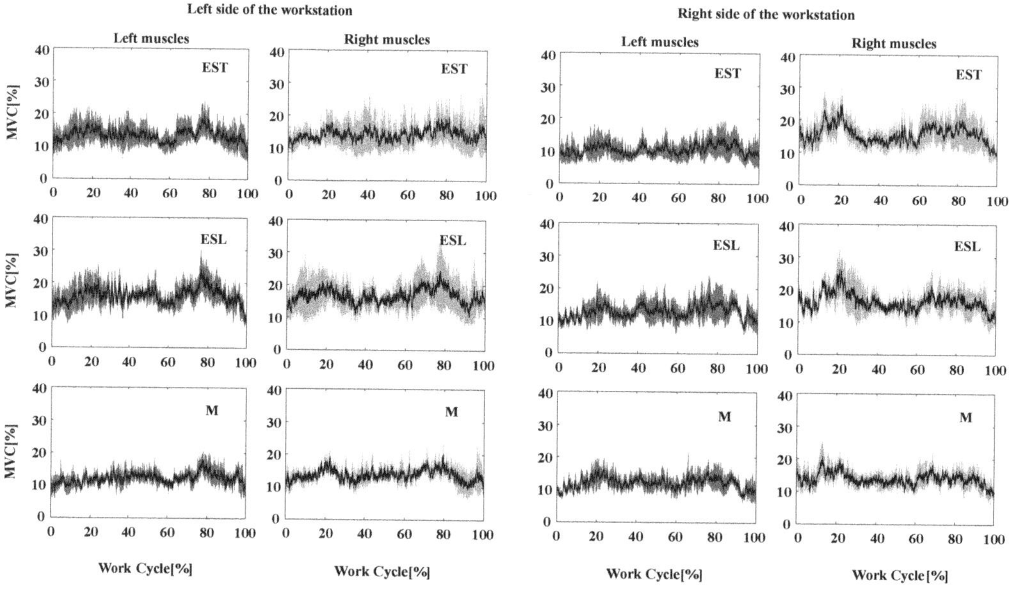

Figure 6. Average values (solid black line) and standard error of the mean (SEM) (light and dark grey coloured areas, left and right sides muscles, respectively) of the muscle activity of paravertebral muscles (Erector Spinae Thoracic (EST), Erector Spinae Lumbar (ESL), and Multifidus (M)) from three workers (W1, W2, and W3) at the left and right workstations, respectively.

Table 1 reported the comparisons between the average values (±SD—standard deviation) of each enrolled worker's muscle activity (left and right paravertebral complex) while performing activities on the left and right side of the workstation.

Table 1. The mean values (±SD) of the muscle activity are expressed as a percentage of MVCs.

	W1		W2		W3	
	Trunk Muscle Side		**Trunk Muscle Side**		**Trunk Muscle Side**	
Right side of the workstation	Left	Right	Left	Right	Left	Right
EST	6.7 ± 1.9	16.9 ± 2.7	8.9 ± 2.4	10.5 ± 2.2	15.7 ± 5	23 ± 3
ESL	10.4 ± 1.7	14.2 ± 3.6	14.6 ± 3.8	22 ± 6	13.7 ± 3.2	16.5 ± 2.8
M	10.1 ± 2	15.5 ± 2.7	12.9 ± 4	9.9 ± 2.2	14 ± 1.4	18.2 ± 2.8
Left side of the workstation	Left	Right	Left	Right	Left	Right
EST	9.5 ± 1.6	Left	17.2 ± 5.3	11.7 ± 2.3	17.2 ± 1.6	23.4 ± 4
ESL	18.5 ± 5.7	12.9 ± 2.8	20.6 ± 4.1	25.2 ± 7.4	13.6 ± 2.4	15.7 ± 2.7
M	14 ± 6.9	13.8 ± 4.3	11.2 ± 2.2	14.6 ± 6.7	14.2 ± 3.4	17.8 ± 4.4

W—worker; EST—Erector Spinae Thoracic region; ESL—Erector Spinae Lumbar region; M—Multifidus.

Figure 7 compares the mean values (±SD) of the muscle activity (right and left paravertebral complex) of all enrolled workers during the work performance on the right and left side of the workstation.

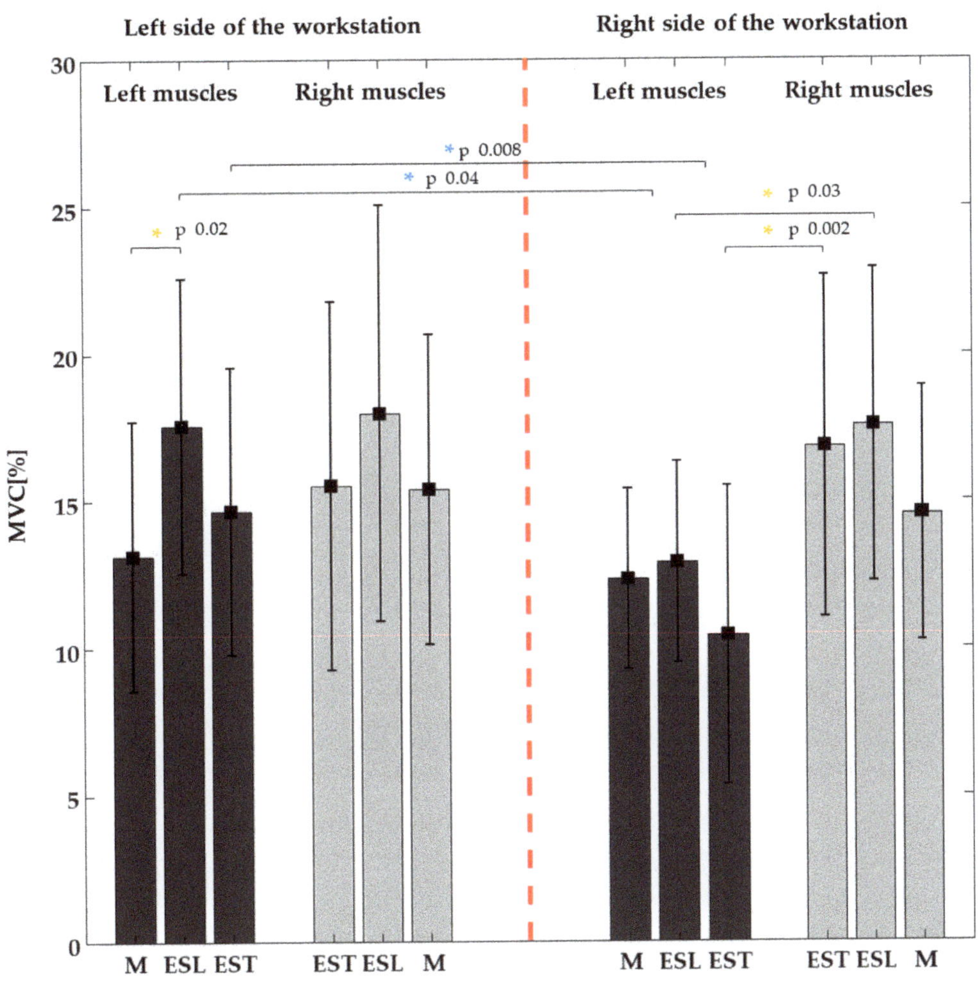

Figure 7. Muscle activity average values of overall paravertebral muscles (M—Multifidus; EST—Erector Spinae Thoracic region; ESL—Erector Spinae Lumbar region). The comparison between the right and left side of the workstation (BLUE stars across the two sides of the workstation) and between the right and left muscles of the same side of the workstation (YELLOW stars on the same side of the workstation) show a statistical difference in the activation among the different muscle groups (* *p* value).

On the right side of the workstation, a significant statistical asymmetry (difference in values between right and left paravertebral muscles) of the muscular activity between the two paravertebral muscle groups EST ($p = 0.002$) and ESL ($p = 0.03$) was reported, but not between muscles group M. In particular, the right-side muscles (16.8 ± 5.8% EST and 17.6 ± 5.3% ESL) show increased activity compared to those on the left side (10.4 ± 5% EST and 12.9 ± 3.4%). This asymmetry is not observed on the workstation's left side. However, on the left-side workstation, in the left paravertebral complex, the muscle M (13.1 ± 4.6%) shows significantly less activity than ESL (17.6 ± 5%) ($p = 0.02$). There are two other significant differences in the results, i.e., between the right and left side of workstations on two left paravertebral complexes EST ($p = 0.008$) and ESL ($p = 0.04$). The EST (14.7 ± 4.9%) and ESL (17.6 ± 5%) muscles on the left side of the workstation present a higher activity than the EST (10.4 ± 5%) and ESL (12.9 ± 3.4%) on the right one.

Finally, from Figure 7, it is possible to observe an expected behaviour, both on the paravertebral complexes of the left and right sides muscles and on both sides of the workstation, i.e., an ever-higher activity of the ESL muscles.

3.2. Kinematic Analysis

The kinematic signals related to trunk flexion forward angle (Figures 8 and 9) were acquired and processed to obtain data shown in Table 2.

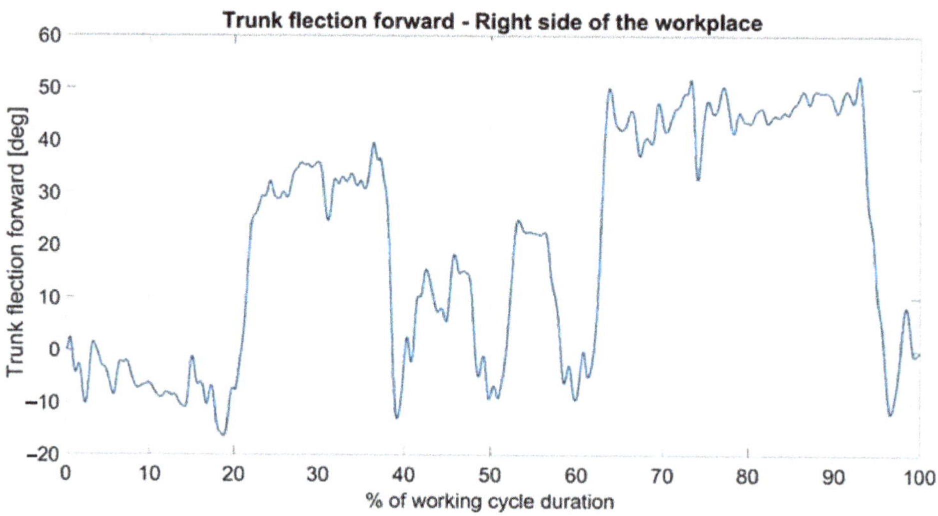

Figure 8. Kinematic signal of trunk flexion forward in one working cycle (right side of the workstation).

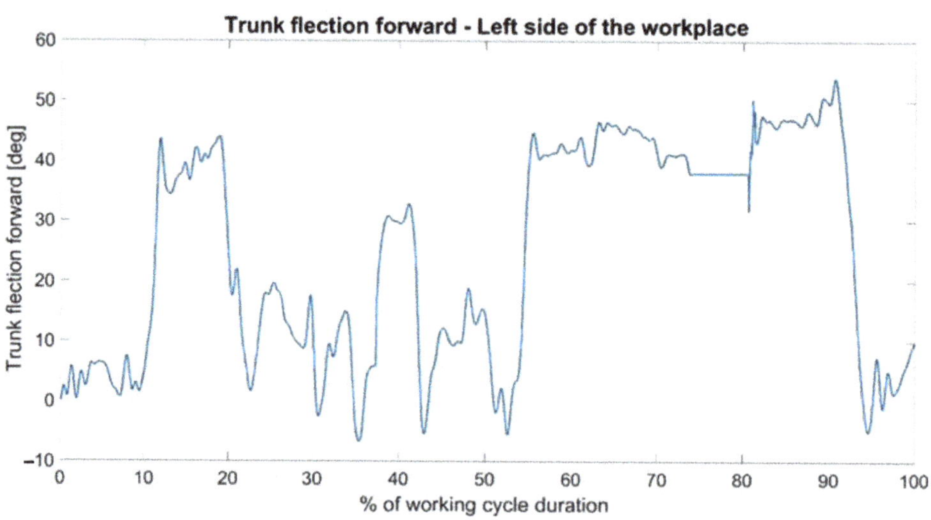

Figure 9. Kinematic signal of trunk flexion forward in one working cycle (left side of the workstation).

Table 2. Trunk flexion forward static posture data analysis: mean and peak flexion forward angle and duration values.

	Reference Ranges	20–60°			>60°		
		W1	W2	W3	W1	W2	W3
Right side of the workstation	Mean value (±SD) [°]	40.63 ± 6.25	39.32 ± 11.35	46.40 ± 4.29	-	77.90 ± 8.23	-
	Peak value [°]	50.71 ± 4.96	54.41 ± 3.74	56.55 ± 2.97	-	86.52 ± 8.60	-
	Mean value (±SD) of posture duration [t]	19.54 ± 2.58	12.00 ± 6.69	22.18 ± 1.71	0	10.82 ± 1.65	0
	Posture duration in % of working cycle [%]	34	21	38	0	19	0
Left side of the workstation	Mean value (±SD) [°]	38.83 ± 5.61	37.26 ± 5.28	43.15 ± 6.00	-	82.77 ± 4.99	65.29 ± 1.42
	Peak value [°]	49.84 ± 7.86	50 ± 6.61	56.52 ± 3.52	-	94.06 ± 6.47	71.01 ± 0.87
	Mean value (±SD) of posture duration [t]	17.32 ± 5.75	11.83 ± 3.40	16.85 ± 3.84	0	17.89 ± 7.15	6.37 ± 2.72
	Posture duration in % of working cycle [%]	30	20	29	0	31	11

From both Figures 8 and 9, it is possible to identify bending peaks of greater or lesser duration linked to specific operations (installation of the cabinet—single arrow—and assembly and wiring operations—double arrow).

Table 2 shows data on kinematic signal processing and refers only to static postures (postures held for at least 4 s consecutively) by considering the angle values within the range 20–60° and higher than 60°, according to ISO 11226. For this reason, postures below 20 degrees are not shown.

The average values (±SD) of the static trunk flexion posture angles recorded during working activities were 43 (±10.1) degrees for the left side and 44 (±6) degrees for the right side during the whole recorded working activity, showing no statistical difference ($p > 0.05$). The average values (±SD) of the total time in the static trunk flexion forward fixed posture recorded during the whole work activity were 21.8 (±4.8) s for the left side of the workstation and 20 (±3.9) s for the right side of the workstation. No statistical difference was found ($p > 0.05$). It is possible to state that the working activities on both sides are well balanced regarding the postural load of the trunk.

4. Discussion

The purpose of this study was to evaluate a combined methodological approach based on the concurrent use of sEMG and a set of inertial sensors for the quantitative risk assessment of biomechanical overload. Three automotive industry workers were enrolled to evaluate the biomechanical effort during the working activity in the assembly line.

The sEMG results suggest a significant involvement of trunk muscles in the studied working task. Indeed, the mean muscle activation values were between 10% and 20% of MVCs, particularly the Erector Spinae Lumbar Region and the Multifidus. According to kinematic data, these values refer to a quite relevant effort of trunk muscles and show a significant biomechanical load at the spine level. Moreover, these muscles have been studied because they have an independent function for stabilisation and are crucial for the stability and mobility of the lumbar spine, determining a main aetiological action for low back pain. The extent of alterations in the structure and muscle function of the paraspinal muscles could be related to the recurrence or chronicisation of low back pain [32]. Therefore, the results suggest that it is important to analyse this kind of work task in the real work environment to gain quantitative measurements to propose and verify ergonomic solutions and changes in work organisation for low back pain and musculoskeletal disorders prevention.

Moreover, sEMG RMS data show a significant difference between the muscle activity of the two body sides, identifying an asymmetry. These results relate to inherent trunk torsional components during dynamic movements when workers follow moving cars in their assembly activity. It is worth underlining that this muscular behaviour agreed with the preliminary observations of the working tasks before the experimental sessions. Further sEMG and kinematic studies will allow a better understanding of how trunk flexion and torsion are combined during tasks. This is important information considering the pathogenetic role of trunk bending and twisting.

About the kinematic motion analysis, the contribution to the biomechanical overload due to the static working posture of the trunk is made mainly by the flexion forward.

We decided to study trunk flexion for several reasons. The main risk factor for the emergence of low back diseases is non-neutral trunk postures, especially those requiring flexion [33]. Yet, over the past three decades, several quantitative techniques—such as electro-goniometers and inclinometers—have been created to obtain precise measurements of trunk postures in real working environments. Unfortunately, some of them require the application of additional external structures to the subject's skin, which is uncomfortable and makes them unsuitable for long-term measurements [34]. In our opinion, sEMG appears to overcome all of these constraints, although additional field study is required to assess the overload of the arms specifically.

Both electromyographic and kinematic analysis results highlight a muscular effort in the various phases of the work cycle. Just comparing the signals (Figure 6 with Figures 8 and 9) shows a higher muscle effort with a concomitant higher value of trunk flexion, from 20 to 40% of the cycle time and from 60 to 90% of the cycle time.

Another important aspect is the non-invasiveness of the methods applied. Furthermore, these techniques were well tolerated by the workers, and they did not interfere with their performance either. Beyond the company's production needs, this aspect is essential for studying the muscles in real working conditions: the working gesture must not be altered by the possible encumbrance of the probes that the worker must minimally perceive. Moreover, it is necessary to verify that there are no physical interferences between the instruments and the working environment, specifically electromagnetic interference. To overcome this issue, in this research, a commercial system of sEMG probes was used; it communicates wirelessly with a device connected to a notebook that acquires and processes the signal. This is a closed system that, in this study, did not show interferences with the instrumentation on the production line, and no systematic signal distortions were observed in the kinematic data processing. Therefore, for this test, the two systems (sEMG and inertial sensors suites) were found to be simultaneously usable, both in terms of wearability and in the absence of interference in the reception of signals.

The results of this study confirm what has emerged from research in the literature, i.e., the contribution of information about the biomechanical overload, which the wearable sensor technology can provide. Another possible advantage could be integrating this information with current risk assessment methods to obtain a risk evaluation [16]. Concerning other research, this study provides data acquired in a working environment during real production. However, the research carried out needs to be improved.

Due to specific technical difficulties that arose during the execution of the surveys and that affected the suite of inertial sensors, it was not possible to focus attention on the movement's lateral bending and torsional components. These technical difficulties have allowed improvements to be made to the software component of the suite.

Another limit is the sample number. The results of the electromyographic and kinematic data refer to only three subjects considered adequate for the type of study conducted (pilot study) but are limited in generalising the data. However, the study's outcome was the applicability of the two methods under real working conditions: the obtained data highlighted the main aspects to be considered when applying a protocol that integrates sEMG and kinematics. Future research must necessarily lead to a study with a more significant number of subjects.

The electromyographic analysis then focused on studying amplitude parameters for quick information about muscle activation. Further studies for developing a validated risk assessment method will have to consider the analysis of fatigue and frequency parameters using specific validated protocols presented in the literature (EVA, JASA) [21].

5. Conclusions

This work presents a wearable wireless system that collects data to assess the biomechanical overload, characterised by the combined use of surface electromyography electrodes and a wearable inertial motion capture system. This approach proved that both methods give helpful information about parameters that can be used to evaluate the biomechanical load due to working postures. Future developments will include studies with a larger sample, further analysis of electromyographic signals, and possible integration with specific observational working methods (such as EAWS for the automotive sector).

Author Contributions: Conceptualization, M.G.L.M., F.D., F.C. and N.M.; methodology, M.G.L.M., F.D., and F.C.; formal analysis, L.F., A.M., A.S. and F.C.; investigation, M.G.L.M., A.M. and L.F.; data curation, M.G.L.M., A.M., L.F., E.M.G. and F.D.; writing—original draft preparation, M.G.L.M., A.M., E.M.G., L.F. and F.D.; writing—review and editing, M.G.L.M., A.M., L.F., F.D., A.S., F.C., M.M., E.M.G., N.M. and M.L.; supervision, M.G.L.M., N.M., F.C. and F.D. All authors have read and agreed to the published version of the manuscript.

Funding: This research received no external funding.

Data Availability Statement: The data presented in this study are available on reasonable request from the first author.

Acknowledgments: The authors acknowledge FCA S.p.A. for its availability for us to access the plant and carry out experimental sessions.

Conflicts of Interest: The authors declare no conflict of interest.

References

1. Schneider, E.; Irastorzav, X. OSH in Figures: Work-Related Musculoskeletal Disorders in the EU—Facts and Figures. European Agency for Safety and Health at Work (EU-OSHA). 2010. Available online: https://osha.europa.eu/en/tools-and-publications/publications/reports/TERO09009ENC (accessed on 31 January 2023).
2. Da Costa, B.R.; Vieira, E.R. Risk factors for work-related musculoskeletal disorders: A systematic review of recent longitudinal studies. *Am. J. Ind. Med.* **2010**, *53*, 285–323. [CrossRef]
3. Punnett, L. Musculoskeletal disorders and occupational exposures: How should we judge the evidence concerning the causal association? *Scand. J. Public. Health* **2014**, *42* (Suppl. S13), 49–58. [CrossRef]
4. Bernard, P.B. *Musculoskeletal Disorders and Workplace Factors: A Critical Review of Epidemiologic Evidence for Work-Related Musculoskeletal Disorders of Neck, Upper Extremity, and Low Back*; National Institute for Occupational Safety and Health: Cincinnati, OH, USA, 1997. Available online: https://www.cdc.gov/niosh/docs/97-141/default.html (accessed on 31 January 2023).
5. Takala, E.P.; Pehkonen, I.; Forsman, M.; Hansson, G.Å.; Mathiassen, S.E. Systematic evaluation of observational methods assessing biomechanical exposures at work. *Scand. J. Work. Environ. Health* **2010**, *36*, 3–24. [CrossRef] [PubMed]
6. Kee, D.; Karwowski, W. A comparison of three observational techniques for assessing postural loads in industry. *Int. J. Occup. Saf. Ergon.* **2007**, *13*, 3–14. [CrossRef] [PubMed]
7. Karhu, O.; Kansi, P.; Kuorinka, I. Correcting working postures in industry: A practical method for analysis. *Appl. Ergon.* **1977**, *8*, 199–201. [CrossRef] [PubMed]
8. Brandl, C.; Mertens, A.; Schlick, C.M. Ergonomic analysis of working postures using OWAS in semi-trailer assembly, applying an individual sampling strategy. *Int. J. Occup. Saf. Ergon.* **2017**, *23*, 110–117. [CrossRef] [PubMed]
9. McAttamney, L.; Corlett, E.N. RULA: A survey method for the investigation of work-related upper limb disorders. *Appl. Ergon.* **1993**, *24*, 91–99. [CrossRef] [PubMed]
10. Hignett, S.; McAtamney, L. Rapid entire body assessment (REBA). *Appl. Ergon.* **2000**, *31*, 201–205. [CrossRef]
11. Schaub, K.; Caragnano, G.; Britzke, B.; Bruder, R. The European Assembly Worksheet. *Theor. Issues Ergon. Sci.* **2013**, *14*, 616–639. [CrossRef]
12. Vieira, E.R.; Kumar, S. Working postures: A literature review. *J. Occup. Rehabil.* **2004**, *14*, 143–159. [CrossRef]
13. Blab, F.; Avci, O.; Daub, U.; Schneider, U. New approaches for analysis in ergonomics: From paper and pencil methods to biomechanical simulation. In *16. Internationales Stuttgarter Symposium*; Bargende, M., Reuss, H.C., Wiedemann, J., Eds.; Springer: Wiesbaden, Germany, 2016. [CrossRef]
14. Battini, D.; Persona, A.; Sgarbossa, F. Innovative real-time system to integrate ergonomic evaluations into warehouse design and management. *Comput. Ind. Eng.* **2014**, *77*, 1–10. [CrossRef]

15. Ferguson, S.A.; Allread, G.W.; Le, P.; Rose, J.; Marras, W.S. Shoulder Muscle Fatigue During Repetetive Tasks as Measured by Electromyography and Near-Infrared Spectroscopy. *Hum. Factors* **2013**, *55*, 1077–1087. [CrossRef] [PubMed]
16. Peppoloni, L.; Filippeschi, A.; Ruffaldi, E.; Avizzano, C.A. (WMSDs issue) A novel wearable system for the online assessment of risk for biomechanical load in repetitive efforts. *Int. J. Ind. Ergon.* **2015**, *37*, 563–571. [CrossRef]
17. Vignais, N.; Miezal, M.; Bleser, G.; Mura, K.; Gorecky, D.; Marin, F. Innovative system for real-time ergonomic feedback in industrial manufacturing. *Appl. Ergon.* **2013**, *44*, 566–574. [CrossRef] [PubMed]
18. Vignais, N.; Bernard, F.; Touvenot, G.; Sagot, J.C. Physical risk factors identification based on body sensor network combined to videotaping. *Appl. Ergon.* **2017**, *65*, 410–417. [CrossRef] [PubMed]
19. Ghasemzadeh, H.; Jafari, R.; Prabhakaran, B. A Body Sensor Network with Electromyogram and Inertial Sensors: Multi-Modal Interpretation of Muscular Activities. *IEEE Trans. Inf. Technol. Biomed.* **2010**, *14*, 198–206. [CrossRef]
20. Ranavolo, A.; Draicchio, F.; Varrecchia, T.; Silvetti, A.; Iavicoli, S. Wearable Monitoring Devices for Biomechanical Risk Assessment at Work.: Current Status and Future Challenges—A Systematic Review. *Int. J. Environ. Res. Public Health* **2018**, *15*, 2001. [CrossRef]
21. Merletti, R.; Farina, D. *Surface Electromyography: Physiology, Engineering and Applications*; IEEE Press/J Wiley: Hoboken, NJ, USA, 2016. [CrossRef]
22. Pigini, L.; Colombini, D.; Rabuffetti, M.; Ferrarin, M. Tecniche di acquisizione ed analisi del segnale elettromiografico per lo studio del sovraccarico biomeccanico occupazionale. *Med. Lav.* **2010**, *101*, 118–133.
23. Draicchio, F.; Silvetti, A.; Ranavolo, A. Il contributo dell'elettromiografia di superficie (sEMG) alla valutazione del rischio biomeccanico nelle attività industriali. *G. Ital. Med. Lav. Erg.* **2011**, *33*, 226–229.
24. Filippeschi, A.; Schmitz, N.; Miezal, M.; Bleser, G.; Ruffaldi, E.; Stricker, D. Survey of Motion Tracking Methods Based on Inertial Sensors: A Focus on Upper Limb Human Motion. *Sensors* **2017**, *17*, 1257. [CrossRef]
25. Caputo, F.; Greco, A.; D'Amato, E.; Notaro, I.; Lo Sardo, M.; Spada, S.; Ghibaudo, L. A Human Postures Inertial Tracking System for Ergonomic Assessments (Conference Paper). In *Proceedings of the 20th Congress of the International Ergonomics Association (IEA 2018), Florence, Italy, 26–30 August 2018*; Bagnara, S., Tartaglia, R., Albolino, S., Alexander, T., Fujita, Y., Eds.; Advances in Intelligent Systems and Computing; Springer: Cham, Switzerland, 2018; Volume 825. [CrossRef]
26. Poitras, I.; Bielmann, M.; Campeau-Lecours, A.; Mercier, C.; Bouyer, L.J.; Roy, J.-S. Validity of Wearable Sensors at the Shoulder Joint: Combining Wireless Electromyography Sensors and Inertial Measurement Units to Perform Physical Workplace Assessments. *Sensors* **2019**, *19*, 1885. [CrossRef] [PubMed]
27. Merino, G.; da Silva, L.; Mattos, D.; Guimarães, B.; Merino, E. Ergonomic evaluation of the musculoskeletal risks in a banana harvesting activity through qualitative and quantitative measures, with emphasis on motion capture (Xsens) and EMG. *Int. J. Ind. Ergon.* **2019**, *69*, 80–89. [CrossRef]
28. Monaco, M.G.L.; Fiori, L.; Marchesi, A.; Greco, A.; Ghibaudo, L.; Spada, S.; Caputo, F.; Miraglia, N.; Draicchio, F. Biomechanical overload evaluation in manufacturing: A novel approach with sEMG and inertial motion capture integration(Conference Paper). In *Proceedings of the 20th Congress of the International Ergonomics Association (IEA 2018), Florence, Italy, 26–30 August 2018*; Bagnara, S., Tartaglia, R., Albolino, S., Alexander, T., Fujita, Y., Eds.; Advances in Intelligent Systems and Computing; Springer: Cham, Switzerland, 2018. [CrossRef]
29. Barbero, M.; Merletti, R.; Rainoldi, A. *Atlas of Muscle Innervation Zones: Understanding Surface Electromyography and Its Applications*; Springer: Milan, Italy, 2012. [CrossRef]
30. Varrecchia, T.; De Marchis, C.; Rinaldi, M.; Draicchio, F.; Serrao, M.; Schmid, M.; Conforto, S.; Ranavolo, A. Lifting activity assessment using surface electromyographic features and neural networks. *Int. J. Ind. Ergon.* **2018**, *66*, 1–9. [CrossRef]
31. Ranaldi, S.; De Marchis, C.; Conforto, S. An automatic, adaptive, information-based algorithm for the extraction of the sEMG envelope. *J. Electromyogr. Kinesiol.* **2018**, *42*, 1–9. [CrossRef]
32. Goubert, D.; De Pauw, R.; Meeus, M.; Willems, T.; Cagnie, B.; Schouppe, S.; Van Oosterwijck, J.; Dhondt, E.; Danneels, L. Lumbar muscle structure and function in chronic versus recurrent low back pain: A cross-sectional study. *Spine J.* **2017**, *17*, 1285–1296. [CrossRef]
33. Wai, E.K.; Roffey, D.M.; Bishop, P.; Kwon, B.K.; Dagenais, S. Causal assessment of occupational bending or twisting and low back pain: Results of a systematic review. *Spine J.* **2010**, *10*, 76–88. [CrossRef]
34. Porta, M.; Pau, M.; Orrù, P.F.; Nussbaum, M.A. Trunk Flexion Monitoring among Warehouse Workers Using a Single Inertial Sensor and the Influence of Different Sampling Durations. *Int. J. Environ. Res. Public Health* **2020**, *17*, 7117. [CrossRef]

machines

MDPI

Article

A Human Error Analysis in Human–Robot Interaction Contexts: Evidence from an Empirical Study

Mario Caterino [1,*], Marta Rinaldi [2], Valentina Di Pasquale [1], Alessandro Greco [2], Salvatore Miranda [1] and Roberto Macchiaroli [2]

1 Department of Industrial Engineering, University of Salerno, 84084 Fisciano, Italy; vdipasquale@unisa.it (V.D.P.); smiranda@unisa.it (S.M.)
2 Department of Engineering, University of Campania Luigi Vanvitelli, 81031 Aversa, Italy; marta.rinaldi@unicampania.it (M.R.); alessandro.greco@unicampania.it (A.G.); roberto.macchiaroli@unicampania.it (R.M.)
* Correspondence: macaterino@unisa.it

Abstract: More than 60 years has passed since the installation of the first robot in an industrial context. Since then, industrial robotics has seen great advancements and, today, robots can collaborate with humans in executing a wide range of working activities. Nevertheless, the impact of robots on human operators has not been deeply investigated. To address this problem, we conducted an empirical study to measure the errors performed by two groups of people performing a working task through a virtual reality (VR) device. A sample of 78 engineering students participated in the experiments. The first group worked with a robot, sharing the same workplace, while the second group worked without the presence of a robot. The number of errors made by the participants was collected and analyzed. Although statistical results show that there are no significant differences between the two groups, qualitative analysis proves that the presence of the robot led to people paying more attention during the execution of the task, but to have a worse learning experience.

Keywords: human–robot interaction; human error analysis; empirical study; virtual reality

1. Introduction

The first application of industrial robotics was in 1961, when General Motors introduced the UNIMATE robot in its factory to perform repetitive and dangerous tasks for human operators [1]. Since then, the lower prices due to technological developments and the need for replacing human workers with machines paved the way for the adoption of robots in many sectors. The introduction of robots into the industry was one of the main innovations during the third industrial revolution, and, today, half of the global manufacturing companies have at least one robot in their factories [2]. One of the main differences between the robots adopted in industrial systems during the third industrial revolution and the ones developed within the fourth industrial revolution is the degree of collaboration with human workers. In the past, human workers could not interact with robots, which were confined in delimited zones for carrying out the programmed tasks, often dangerous for humans because of the execution speed needed for the series of productions. In recent years, the rise of new technologies increased the safety of machines, including robots. Thus, the concept of human–robot interaction (HRI) has arisen and seems to be very promising for improving the performance of industrial systems, while also considering the social aspects related to human jobs, which may become less repetitive and exhausting. The HRI discipline brings together scholars and practitioners from various domains (engineers, psychologists, designers, etc.) to study the best solutions for integrating robots that can interact with humans and the social impacts of such an interaction [3]. The common opinion in the scientific community is that human behavior and, consequently, the performance of humans in industrial environments, is strongly affected by the feelings of trust and safety toward robots [4].

Safety is a crucial aspect of HRI because the closeness to the robots can cause injuries to humans due to excessive energy/power transferred to the robots [5]. Zachakari et al. [6] and Sharkawy and Koustoumpardis [7] claimed that the safety problem in HRI needs to be addressed from different perspectives, in relation to the design of the robots, the standards, and human psychology. The design of robots in HRI concerns the implementation of control methods to tune the speed and the force elapsed by robots according to their proximity to a human worker. In Caterino et al.'s work [8], the authors presented an industrial application in which algorithms were introduced as control methods for the speed of the robot according to the proximity to the worker. The method is based on the standard ISO/TS 15066 [9], which provides the guidelines for safety in HRI by defining four different degrees of collaboration: (i) safety-rated monitored stop, when a robot ceases its motion because a worker is entering in the collaborative workspace; (ii) hand guiding, when workers use hand-operated devices to transmit motion command to the robot; (iii) speed and separation monitoring, when robot systems and workers may move concurrently in the shared workspace (co-existence); and (iv) power and force limiting, when physical contacts between the workers and the robots can occur either intentionally or unintentionally.

It is clear that, according to the degree of HRI, the psychological aspect alongside the trust of the human worker in the robot system plays a significant role. Trust is "the attitude that an agent will help achieve an individual's goals in a situation characterized by uncertainty and vulnerability" [10]. In HRI contexts, trust is important because it affects human decisions and behavior. Social studies have demonstrated that the trust in robots by individuals is linked to their personal previous experiences, attitudes, and the risks connected to the execution of the tasks [11,12]. In industrial contexts, trust has been assessed by Charalambous et al. [13], who observed that it mainly depends on the robot's motion and speed and on the safety level perceived by the human during collaboration.

According to the level of trust, safety, and the design of robots, human performance can be affected while executing a task in collaborative workplaces and can lead to making errors, which can, at times, result in serious injuries, but, of course, is always an indicator of the human performance within industrial environments, especially in HRI cases [14].

This paper aims to investigate such an aspect related to human performance in an industrial environment, i.e., the number of errors made during the working activity. To achieve this scope, the errors made by two groups of workers, one working in a shared environment with robots and one working without, have been assessed. The assessments have been performed through the use of VR, and statistical analyses have been carried out to evaluate if significant differences in the errors made by the two groups of workers were verified. Differences would lead to conclude that the presence of robots may affect human performance.

The use of VR tools falls within the scope of promoting or testing efficient and safe industrial systems and improving the worker's well-being [15,16] through digital tools developed during the era of the fourth industrial revolution. In HRI context, the use of digital tools represents a strategic way to enable dynamic and safe analyses. Among several tools, VR reproduces the working environment, simulates the production system, and allows for the quick, safe, and economical interaction between humans and robots [17]. For years, the use of VR has been validated as an effective tool to simulate industrial HRI systems [18], and it has demonstrated that it can provide performance assessments of HRI systems [19], including errors [14].

The remainder of the paper is presented as follows. Section 2 analyzes the literature related to the scope of this paper. In Section 3, the experimental phase is explained, while in Section 4, the results are reported and discussed. Finally, Section 5 reports the main conclusions.

2. Literature Review

Human errors in working environments can occur. The consequences of such errors can range from wasted time to more serious problems, affecting the safety of the workers themselves, especially in high-risk fields such as nuclear plants or aerospace [20]. In all cases, human errors represent financial losses for the companies. For such a reason, the study of human errors has become increasingly important, and today, it is a well-defined discipline called the human reliability analysis (HRA).

HRA is the discipline that aims to assess the reliability of a complex system, in which humans and machines work together. To elaborate, HRA evaluates the worker's contribution to the reliability of the system, trying to predict human error rates and understanding the impact caused by human errors in such systems [21]. Although each method attempting to evaluate the human error probability (HEP) falls under the umbrella of HRA, it is commonly recognized by scientists that HRA has three main functions [22]: (i) the identification of human errors, (ii) the prediction of the likelihood of occurrence; and (iii) the reduction in the probability of the occurrence of errors. The focus of this paper is on (i) and (ii).

HRA can be conducted by utilizing several methods [23] and its domain of application is very wide. In the assembly and manufacturing fields, HRA methods have been used for different applications and may have different purposes. Kern and Rafflinghaus developed an HRA for manual assembly operations in the automotive sector, demonstrating its usefulness in developing time-optimized and quality-optimized assembly operations [24]. The same authors also demonstrated that implementing systematic HRA methods provides companies with a reliable tool for error risk prediction [25]. Caputo et al. [26] defined an HRA model that, besides mapping error types and analyzing the logic conditions in which the errors occurred, was also able to identify the errors causing the highest economic impact and to provide a quantitative assessment of the economic impact of the errors. Di Pasquale et al. used an HRA method to simulate the optimal breaks during a work shift in a manual assembly line [27]. Concerning the causes of errors, Torres et al. highlighted that, during the execution of manual assembly tasks, some operations are the most likely to cause errors because the level of difficulty is very high [28]. In other cases, the main reason for the errors can be related to the poor skills of the operators or the adverse psychology, often due to the scarce confidence in the working environment or the lack of safety awareness [29]. The psychological factors which ledd to human errors have been widely debated in the literature, beginning from 1990, when the study of HRA started combining different areas of scientific studies, such as behavioral science and psychology [30]. The need to also study the psychological aspect of HRA arose when it became evident that human behavior was impacted by many aspects that do not impact a machine. The working environment, the mental and physiological factors, and many others have been widely recognized as possible causes of human errors [31]. In this context, it is not possible to neglect these factors.

When robots are introduced in shared working spaces in industrial contexts with humans, they represent agents affecting the behavior of human operators. In the case of HRI systems, robots are often seen as "co-workers" by operators, generating social interactions [32]. Additionally, the introduction of cooperative or collaborative robotics leads to many benefits for operators, mainly related to the reduction in efforts due to repetitive and heavy tasks [33]. In other cases, robots may create feelings of fear, anxiety, and surprise, especially when they have dangerous end effectors or move in an unpredictable trajectory [34]. According to Lu et al. [34], the factors generating these feelings are mainly related to two aspects. (i) The characteristics of the robots, such as the dimensions or the speed, which have a connection with the safety awareness. In fact, robots' speed and dimensions lead operators to emphasize their perception, comprehension, and projection of the safety-related events during the daily task to understand the potential hazard of the working environment surrounding them [35]. (ii) Robots' trajectories, which are often felt as unpredictable, confuse the human operator. In fact, in manufacturing contexts, robots are often programmed for minimizing the execution times of a task, thus resulting in unpredictable movement trajectories for the operator working with or near the robots [36].

The presence of the safety awareness issue was also confirmed by a study by Di Pasquale et al. [37], who found that, besides safety, the presence of a robot in a shared working environment also influences ergonomics and productivity.

From the previous literature, it is evident that the human operators' performance can be impacted by HRI. The evaluation of such impacts is carried out by means of different methods, both direct (such as questionnaire [38,39]) and indirect (cardiac response, facial expression, electromyogram, etc. [34]). Some empirical studies used direct and indirect methods together to evaluate the impact of robots on human actions in HRI. The findings are often conflicting. Huber et al. [40] demonstrated that robots influence the behavior of humans in simple tasks as well, such as the ones implying hand movements. Similar results were found by Zhang et al. [41], whose empirical study demonstrated that robots affect humans in some actions, influencing the completion times and trajectories followed. The same results are confirmed by Chen et al. [42], who observed that people changed behavior according to their anticipation of a potential collision with robots during pedestrian motion. On the other hand, the practical findings of the study proposed by Xie et al. [43] highlighted that there is no difference in the behavior of humans when comparing human–human and human–robot activities, even if the participants declare a higher trust in human–human than human–robot activities. Vassallo et al. [44] also observed similar results for pedestrian activity.

Among the indirect methods, operators' error evaluation during the working tasks can be used as an approach to evaluate HRA in HRI [45,46]. Errors directly influence the productivity of the company because they generate defective items. Morioka and Sakakibara demonstrated that HRI reduces the number of defective items thanks to the precision and repeatability of the robots [47]. As opposed to this, the feelings of anxiety generated by the presence of robots may lead to errors by the operators [48]. Thus, there is, again, a conflicting perspective on the impacts of robots on human behavior and performance. To address this problem, this paper aims to provide insights from an experimental study by comparing the errors made by operators working with and without robots in a cooperative industrial environment.

To the best of the authors' knowledge, there are no experimental studies aiming to assess the impact of HRI on the number of errors made by the operators. To cover this research gap, an experimental campaign has been developed and conducted through VR systems to reach the objective of the paper.

3. Materials and Methods

An experimental campaign has been conducted for assessing the impact of robots' presence on human performance by evaluating the number of errors made by the operators in an HRI environment. The experimental campaign has been conducted in a laboratory environment by virtually reproducing a real task of a company working in the aerospace sector.

3.1. Task and Virtual Environment

VR has been used for reproducing the working environment and simulating the task. The task reproduced is an aircraft fuselage panel assembly task and it consists of the application of sealant on a stringer. The task includes several operations to be completed, which involve the use of different tools. For privacy reasons, it is not possible to describe the operations in detail, but a general overview is given in Figure 1, where the main operations of the entire task are represented in the VR environment, and in Table A1 in Appendix A, where the errors made during the task are listed. It is worth noting that the task simulated is the reproduction of a real industrial task and it was not properly created for pursuing the objectives of this paper.

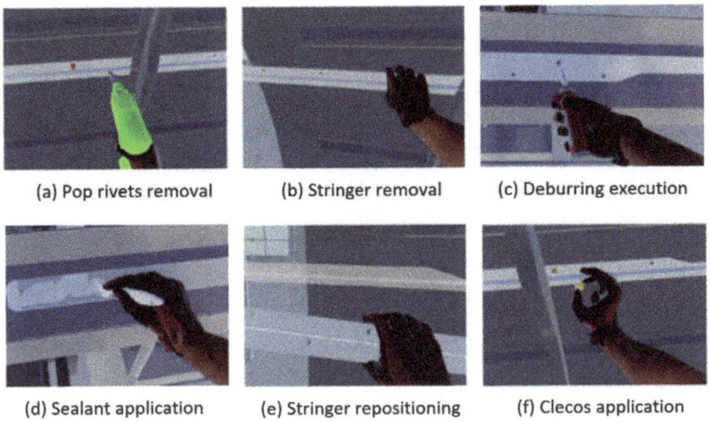

Figure 1. Representation of the task in the VR environment.

For realizing the virtual simulation, the HTC Vive Pro® system was used, while the software used to build the interactions among objects and human players was Unity®. Figure 2 shows the simulated and real environments.

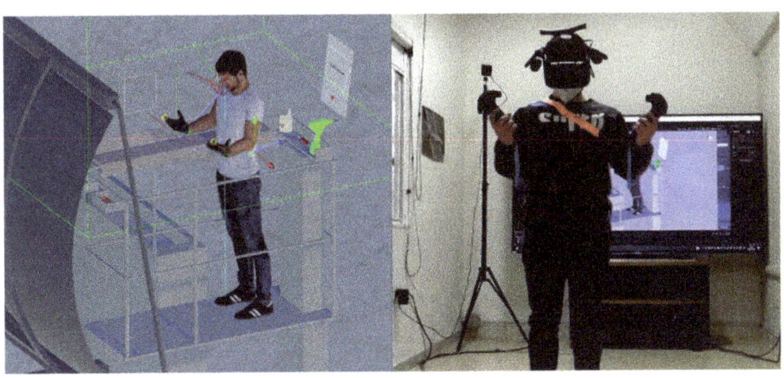

Figure 2. Representation of the virtual and real environments.

To guide the participants through the execution of the required task, a panel was created in the simulated working environment to present instructions on the specific activities to perform.

Two different scenarios have been simulated: scenario#1, in which people perform the task in a shared environment while the robot performs another operation, and scenario#2, in which people perform the task without the presence of the robot. The task that the operator has to perform is identical in both the scenarios. Co-existence has been tested in scenario#1, by considering speed and separation monitoring according to the ISO-TS 15066. Co-existence is a type of HRI where humans and robots share the same workspace, but they cannot have physical contact. When the robot approaches the operator, its speed is reduced until stopping when the distance is under a safety threshold. Figure 3 shows a point of view shot of the operator while completing one of his/her tasks. The presence of the robot was noted at different distances from the operator. The robot used for the experiments is the digital reproduction of the one used in the real industrial environment. The model is a Fanuc M20iA.

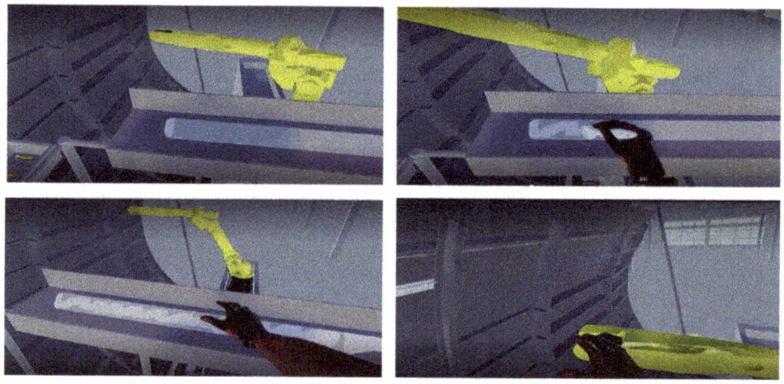

Figure 3. Point of view shot of the robot in the simulated environment.

3.2. Participants

Students from the department of engineering (mainly industrial, aerospace, or mechanical engineering) were recruited for the experiment. Overall, 78 students (33 females, 45 males) volunteered for the experimentation (age, mean = 24.8 years, standard deviation = 3.49 years). The number of students participating in the experiment was high in order to limit some undesired effects, such as the effect of external factors, for example, stress, which may influence the efficiency of the single participant [49].

Participants reported to have no previous experience with robots and have no or very little experience with experiments in virtual environments. However, most participants reported slight familiarity with joystick devices, which were utilized to control the virtual scene. Additionally, no one reported previous experience or knowledge of the task to be performed. None declared to have mobility, sight, or hearing problems.

3.3. Procedure and Experimental Design

The experimental campaign was conducted in a laboratory environment and it was carried out following the rules of the Declaration of Helsinki of 1975 (https://www.wma.net/what-we-do/medical-ethics/declaration-of-helsinki/, (accessed on 10 May 2023)), revised in 2013.

Upon arrival at the laboratory, the context and the setting of the experimental test were introduced, and participants were asked to watch a short video, which trained them on how to interact with VR and how to perform the task (tools that he/she must use and objects of the virtual scene). After that, they were given a copy of the information sheet and they were asked to sign the consent form before starting the experiment. To prevent any potential bias, the purpose of the experiment was not revealed to them during the introduction, and participants were simply told that they would be enrolled in a research study involving VR.

Then, each participant was prepared to interact with VR and assigned to one group (HRI_coex and No_HRI); assignation was random and groups were only balanced by sex. Participants were not informed on which group they belonged; thus, they were not aware of the presence of the robot.

A between-group experiment was conducted so that each person was only exposed to a single condition: group#1 (called "HRI_coex") performed the assembly task with the presence of the robot (scenario#1), and group#2 (called "No_HRI") performed the assembly task without the presence of the robot (scenario#2).

Both groups consisted of 39 participants, which were more than sufficient for reaching statistical significance [50,51]. The characteristics of the groups are reported in Table 1.

Table 1. Characteristics of the two groups.

	HRI_coex	No_HRI
Number of male participants	23	22
Number of female participants	16	17
Mean age of participants (years)	24.5	25.1
Standard deviation of age (years)	3.49	3.49

Experimental conditions were the same across participants. Each participant repeated the task four times without breaks between repetitions. Overall, the test lasted 20 min on average.

After completing the experiment, participants were asked to indicate their opinion on the complexity of the task and their perception on the impact of co-existence on their performance. They were asked to respond to seven questions: six were related to the complexity of the task focused on the specific operations performed during the task, and one was on the perception of the participant about errors made due to the presence of the robot. This last command was only posed to the HRI_coex group. For all the commands, the respondents were asked to indicate their opinion using a 10-point bipolar scale [52], where 1 = "not complex" or "strongly disagree" and 10 = "highly complex" or "strongly agree". The detailed questions can be found in Table 6.

3.4. Error Assessment and Data Collection

The purpose of this paper is to investigate if the number of errors made during the working tasks is affected by the presence of a robot in a co-existing working task. The errors were then classified into different categories according to the difficulty in performing the task and their consequences on the product quality and lead time. Task difficulty was considered since the most difficult operations were those that could generate errors and compromise the final quality of the product, thus generating economic damage as well. Moreover, each error impacts the lead times of the entire task, thereby also producing negative economic consequences. Errors leading to the highest economic consequences are also the ones generating the highest increment in the lead times. Based on these aspects, a panel of experts working in the aerospace field was involved in a roundtable discussion to identify the kind of errors to investigate. Three categories have been proposed and are listed below:

- Low-severity errors: Errors considered in this category cause a small increase in the lead times of the entire task, but do not cause any problem to the final quality of the product.
- Medium-severity errors: Errors considered in this category cause a medium increase in the lead times of the entire task with very limited consequences on the quality of the final product.
- High-severity errors: Errors considered in this category cause an increase in the lead times of the entire task as well as quality damage that compromise the final quality of the products. Products subject to this type of error should be reworked to make their quality conform to the quality standards.

Twelve types of errors have been identified during the experimental study, including the possibility of not performing any error. Each error is briefly described in Table A1 reported in Appendix A, where errors are coded.

According to this classification, data were collected through real-time direct observations and validated through the videos of the simulated environment recorded during the experiments. The errors were manually assessed by expert analysts (researchers in the operations management field).

3.5. Statistical Analysis

The application of the experimental procedure allowed us to obtain independent observations (data) about the number of errors made by each participant. Such data were primarily used to provide an overview of the results, using descriptive statistics in the form of frequencies. Statistical analyses were then used, whenever appropriate, to check the statistical differences between groups. The first assessment was to verify if data had a normal distribution, which is a requirement for the choice of the test to perform. For such reasons, a Shapiro–Wilk test was carried out. The next analysis focused on assessing the statistical differences between the two groups, considering the number of errors made in the single experiments, with a focus on potential differences between males and females. Finally, the differences between the groups considering only the first and last experiments were statistically evaluated to determine if there is a reduction in the number of errors due to the repetition of the experiments. For such assessments, two-tailed Wilcoxon rank sum tests were carried out based on the results of the Shapiro–Wilk test. All the statistical analyses were conducted using R studio® 2023.

4. Results and Discussion

4.1. Descriptive Statistics

As mentioned previously, the participants were divided into two groups, namely the group performing the task in scenario#1 with the robot (group #1, "HRI_coex") and the group in scenario#2, without HRI (group #2, "no_HRI").

Firstly, a qualitative analysis of the errors was performed to highlight the frequency of errors on the total number of operations performed. Figure 4 shows the total frequencies of the errors for the two groups, i.e., the errors made by the two groups in all the experiments.

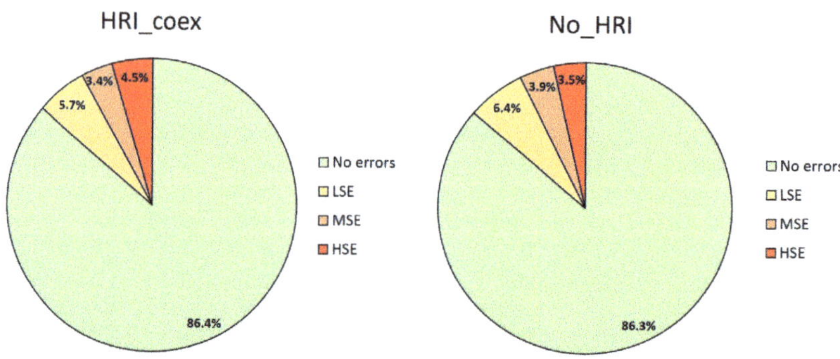

Figure 4. Frequencies of the errors for the groups.

From this first analysis, the reader can observe that the difference on the number of errors committed is only 0.1% between the two groups (13.6% for HRI_coex group and 13.7% for No_HRI group). A small difference can be found in the type of errors made. In particular, the HRI_coex group has a higher percentage of high-severity errors.

The following Table 2 shows the frequency of the total errors made by the two groups for each of the four experiments.

Additionally, by analyzing the results of Table 2, the difference between the two groups seems to be negligible. One remarkable finding is in the difference between the first and second experiments. In fact, although the HRI_coex group performed less errors during the first experiment, the situation was overturned in the second experiment.

Table 2. Characteristics of the two groups.

	HRI_coex	No_HRI
First experiment		
Number of operations without errors	191	186
Number of operations with low-severity errors	19	28
Number of operations with medium-severity errors	19	19
Number of operations with high-severity errors	11	7
Second experiment		
Number of operations without errors	201	211
Number of operations with low-severity errors	15	12
Number of operations with medium-severity errors	11	7
Number of operations with high-severity errors	13	10
Third experiment		
Number of operations without errors	210	210
Number of operations with low-severity errors	17	13
Number of operations with medium-severity errors	5	8
Number of operations with high-severity errors	8	9
Fourth experiment		
Number of operations without errors	217	219
Number of operations with low-severity errors	4	8
Number of operations with medium-severity errors	8	5
Number of operations with high-severity errors	11	8

4.2. Statistical Results

Hypothesis tests were performed to evaluate if there are significant differences in the errors made by the two groups. In order to choose the most appropriate hypothesis test, the normality of the data must be verified. For such a reason, the Shapiro–Wilk test was performed. The test demonstrated that the errors are not normally distributed. This can also be qualitatively noted in Figure 5, where the histograms representing the mean number of errors (x-axis) made by each participant (y-axis) of the two groups in the four experiments are represented and compared with the expected normal curve, given the means and the standard deviations of the groups.

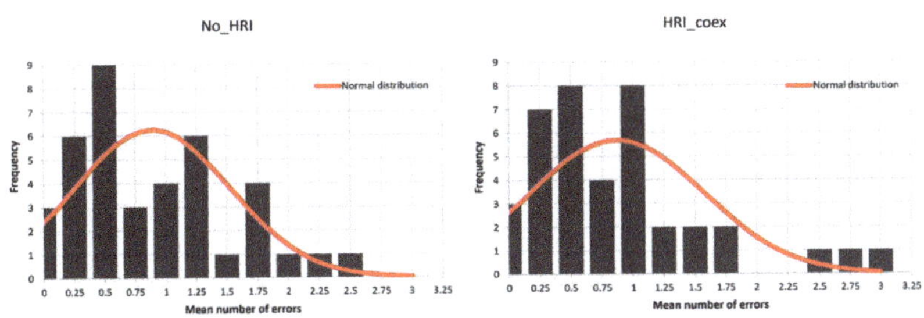

Figure 5. Histograms of the errors of the groups.

Based on the results of the Shapiro–Wilk test, a two-tailed Wilcoxon rank sum test was performed to evaluate if there are significant differences between the two groups, considering both the errors made in the single experiments and the mean of the errors made during the four experiments (Experiment_mean). The Wilcoxon rank sum test is a nonparametric test that can be used to assess if independent samples of data come from the same distribution, even if not normal [53]. In detail, the Wilcoxon rank sum test was used to evaluate if the medians of the two groups come from the same distribution The null and alternative hypotheses, in this paper, are formulated as follows:

- H0: the median of HRI_coex is equal to the median of No_HRI;
- H1: the median of HRI_coex is not equal to the median of No_HRI.

Table 3 reports the values of the tests for the four experiments.

Table 3. Results of the Wilcoxon rank sum test.

	Wilcoxon Statistic	*p*-Value	α-Value
Experiment 1	729.5	0.482	0.05
Experiment 2	866.5	0.497	0.05
Experiment 3	852	0.913	0.05
Experiment 4	783	0.858	0.05
Experiment_mean	765.5	0.741	0.05

The output of the Wilcoxon rank sum test is the same for all the experiments. The *p*-value is always higher than the α-value; thus, it is not possible to reject the null hypothesis. There is no evidence of a significative difference in the medians of the two groups when considering both the single experiments and the mean of the errors.

This can also be observed in Figure 6, where the boxplots of the two groups are shown.

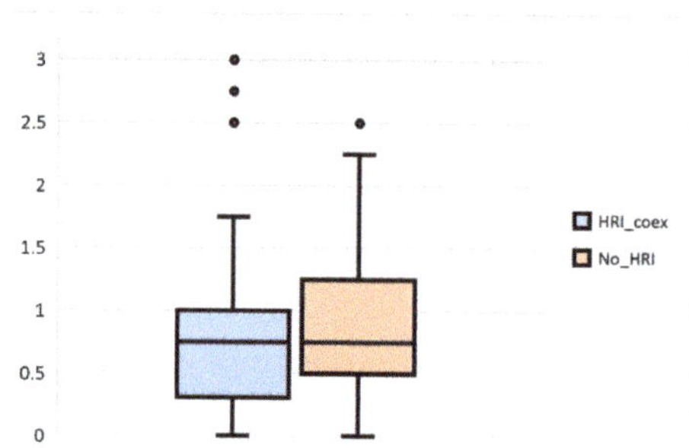

Figure 6. Boxplots of the two groups.

The boxplots show similar characteristics for the two groups, even if the HRI_coex group shows a number of outliers higher than the No_HRI group, which, instead, presents a higher spread of data. A feasible interpretation of this fact may be that the people of the HRI_coex group seem to pay more attention in the tasks' execution, but the effect of the presence of the robot has some kind of impact, generating more errors (three outliers, represented by circles in Figure 6) in some cases.

The second analysis performed was aimed to evaluate if there were significant differences between the male and female sexes related to the two groups. Again, a Wilcoxon rank sum test was performed. The hypotheses made were the same H0 and H1 of the previous analysis. The results are reported in Table 4.

Table 4. Results of the Wilcoxon rank sum test for males and females.

	MALES		FEMALES		
	Wilcoxon Statistic	*p*-Value	Wilcoxon Statistic	*p*-Value	α-Value
Experiment 1	198.5	0.309	162.5	0.947	0.05
Experiment 2	247.0	0.871	194.5	0.236	0.05
Experiment 3	238.5	0.980	140	0.482	0.05
Experiment 4	196	0.261	181	0.433	0.05
Experiment_mean	198.5	0.327	173	0.688	0.05

In all the experiments considered, it is not possible to reject the null hypothesis. Thus, it is possible to affirm that in all the cases analyzed, the differences in the number of errors between the groups (HRI_coex, No_HRI) occurred by chance alone.

The analysis of the boxplots for the males and females (Figure 7) allows highlighting some differences between them.

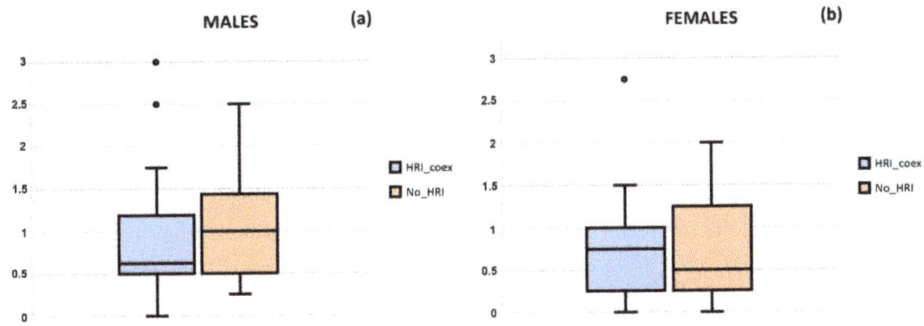

Figure 7. Boxplots of the males (**a**) and females (**b**) errors for the two groups.

For both sexes, the HRI_coex group shows a lower variability than the other. When considering only males (Figure 7a), participants performing the tests with the robot show a median very close to the first quartile. This means that the variance of the errors performed is higher than the median, as opposed to the females (Figure 7b), whose median is closer to the third quartile than the first. In addition to this, it is possible to note that the first and third quartiles are lower for the female (Figure 7b) than males (Figure 7a) and that males present more outliers than the females. Thus, it could be concluded that females seem to pay more attention to task execution than males in case of the presence of robots (HRI_coex). This trend is also confirmed in the case of the No_HRI group, where males show a higher median and higher percentiles than females, even if, in this case, this phenomenon is less evident because of the higher variability of the data highlighted by the boxplots. Finally, by comparing the boxplots of the two groups for the same sex, it is evident that the number of errors has a greater variability for the No_HRI group than the HRI_coex group for both sexes (larger boxes and higher whiskers). Then, once again, this aspect seems to highlight a higher attention paid in the tasks execution by the HRI_coex group.

A Wilcoxon rank sum test was also performed to compare the two groups on a specific activity during the task. The activity is related to the deburring of the panel and it is the only activity in which the participant was oriented with the eyes in the direction of the robot. Although the total number of errors during the four experiments was greater for the HRI_coex group (43 vs. 37), the Wilcoxon rank-sum test does not allow rejecting the null hypothesis (W statistic= 851, p-value = 0.602, α-value = 0.05). Additionally, in this case, it is possible to affirm that the presence of the robot does not significantly affect the number of errors made by the participants.

The last statistical analysis carried out was aimed at evaluating If there are significant differences within the same group on the number of errors made in the first and last experiments. A Wilcoxon rank sum test was performed again, this time with the assumption that a learning phenomenon applies to both groups and, consequently, that the number of errors made in the first and last experiments would be significantly different. The results of the tests are shown in Table 5.

Table 5. Results of the Wilcoxon rank sum test for evaluating differences between the first and fourth experiments.

	Wilcoxon Statistic	p-Value	α-Value
HRI_coex	1108.5	0.0016	0.05
No_HRI	1151.0	0.0003	0.05

For both groups, the p-value is lower than the α-value, indicating that the null hypothesis should be rejected and the difference in the medians is significant between the first and fourth experiments. Thus, for both groups, a learning phenomenon is verified, even if the higher p-value of the HRI_coex may lead to the conclusion that the HRI_coex group had a learning phase less marked than the No_HRI group. This thesis is also supported by the results reported in Table 2, which were determined by observing the number of errors made by the groups in the first and second experiments, whereby, even though the HRI_coex group performed less errors during the first experiment, the situation was overturned in the following experiments, thus confirming that the learning phase for the HRI_coex group is less marked than the No_HRI group.

An assessment was finally carried out to qualitatively evaluate the perception of participants regarding the complexity of the tasks performed and the perceived impact of the robot. Table 6 reports the questions submitted to the participants, while the results of this assessment are reported in Table 7.

Table 6. Questions submitted to the participants after the experiments.

Questions
#1 Express the complexity level of the operation "pop rivets removal".
#2 Express the complexity level of the operation "stringer removal".
#3 Express the complexity level of the operation "deburring execution".
#4 Express the complexity level of the operation "sealant application".
#5 Express the complexity level of the operation "stringer positioning".
#6 Express the complexity level of the operation "positioning the elements supporting the stringer".
#7 Express how much you agree with the following sentence: "the presence of robot leads making a high number of errors".

Table 7. Results of the assessment on the complexity of the task and the perception of robot influence on the errors.

Question No.	Score Range	Results			
		Mean		Standard Deviation	
		HRI_coex	No-HRI	HRI_coex	No-HRI
#1	0 (not complex) 10 (highly complex)	0.76	0.74	2.21	1.96
#2	0 (not complex) 10 (highly complex)	2.57	2.71	2.16	2.37
#3	0 (not complex) 10 (highly complex)	1.28	1.23	2.23	2.33
#4	0 (not complex) 10 (highly complex)	0.73	0.95	2.13	2.30
#5	0 (not complex) 10 (highly complex)	1.97	1.95	2.34	2.53
#6	0 (not complex) 10 (highly complex)	2.07	1.38	2.63	2.26
#7	0 (strongly disagree) 10 (strongly agree)	1.90	NA	2.11	NA

Firstly, it can be observed that both groups had a similar perception on the complexity of the single operations within the task. Thus, it can be concluded that the robot does not have any influence on the perception of the complexity, as expected. According to the participants, all the operations have a low degree of complexity. This confirms the results of Table 2, which highlighted that the majority of the operations were carried out without making any error. The operation of deburring (Question#3), which was the only one with a high severity level and the one which recorded the highest number of errors, was considered not complex, highlighting that participants had no real perception of the errors made during the experiment. Moreover, the participants of the HRI_coex group strongly disagree on the fact that the presence of a robot leads to making a higher number of errors (Question#7). This is a further confirmation of the previous statistical analysis, which highlighted that there are no significant differences between the groups in the number of errors performed.

4.3. Discussion

The results achieved by this case study contribute to the theory in different ways. It was demonstrated that the presence of robots does not generate statistical differences in the human performance. In fact, no significant difference was observed in the number of errors made between the group working in a co-existing environment with robots and the group working without the robot in any of the statistical tests carried out. This is in line with a part of the current literature, such as works by Xie et al. [43] and Vassallo et al. [44], who noticed no difference in HRI and non-HRI activities. Most of the literature argued that an influence is created by the presence of the robot in a working environment [40–42]. The difference between the results of the present study and the other contrasting studies can be explained by the monitored parameters in the papers. In fact, the parameter monitored in the present study is the total number of errors made by two groups and it was empirically observed by observing the real-time experiments and the recorded video. In contrast with the study of Huber et al. [40], who used a very precise motion-tracking system that was able to detect very small differences between participant groups, and that of other papers, which used very sophisticated devices capable to detect very small differences between the monitored physiological parameters of participant groups [34], this paper evaluated a macroscopical parameter (the number of errors) to assess the impact of the presence of the robot on human performance during working tasks. For the reader, it will be easy to understand that monitoring very precise parameters allowed us to better evaluate possible differences between the groups. Our statistical analysis was also confirmed by the participants' answers to a small questionnaire that was submitted to them after the experiments. The qualitative analyses of the answers highlighted that the perception of the complexity of the operations was not affected by the presence of the robot. Moreover, the participants of the HRI_coex group declared that they did not even perceive the robot as a source of nuisance that may have led to increase the number of errors made during the task. However, some small differences between the behaviors of the two groups are also hypothesized in this paper based on some general results. By analyzing the boxplots (Figures 6 and 7), the smaller height of the boxes and the greater number of outliers for the HRI_coex group seem to indicate that greater attention was involved in carrying out the experiments by this group. This could be explained by the presence of the robot, which has the impact of creating feelings of anxiety or stress, making people more focused on the task they are executing. This is in line with what many authors affirmed on the mental stress caused by HRI. The review by Lu et al. [34] demonstrated that psychological factors may affect the behavior and the performance of human operators working in HRI environments. A further demonstration of this is given by the last analysis performed in the previous section, aimed to demonstrate that learning experiences are different in the two groups analyzed. The results showed that, although both groups had a learning experience during the repetition of the experiments, the learning experience of the HRI_coex group was less marked than the No_HRI group. Thus, once again, it seems that the presence of the robot has an impact on the feelings of the participants in the experiments.

5. Conclusions

The present study aimed to evaluate the influence of the presence of a robot in a HRI task performed in an industrial environment. An empirical assessment was carried out for this purpose. A task regarding the assembly of aeronautical parts was simulated in a VR environment and participants were asked to complete the task. Half of the participants performed the test in a co-existing environment with a robot (HRI_coex), half without the robot (No_HRI). The parameter monitored to assess the differences between the groups was the number of errors performed during the execution of the task. The results of the experiment highlight that there are no statistically significant differences between the two groups, even if some considerations can be concluded by investigating the results: the HRI_coex group seems to have been more focused and to have paid more attention to the correct execution of the task. Similar results were observed also by considering the results of male and female participants for the two groups.

This paper has some limitations. Only the number of errors was used as the monitored parameter to assess differences between the groups. It was not considered to evaluate the feelings of the people, which can be assessed through a questionnaire, or the physiological factors by means of specific devices measuring, for example, oxygen intake or the heart rate. Moreover, parameters related to the task, such as the time needed by each participant to accomplish the single experiment, were not considered in this paper. This parameter may be very important because it can provide significant insights on the learning rate for HRI tasks in a virtual environment. In fact, by analyzing the different times employed by participants for each repetition of the task, the learning curves for this type of task could be retrieved. Future developments of this paper will aim to cover these gaps. Moreover, this paper does not consider collaboration, but only co-existence. Future works will consider cooperative tasks to assess the effects of robots on human performance during cooperation, where a physical contact between human and robot is possible.

Author Contributions: Conceptualization, M.C., M.R., V.D.P. and A.G.; methodology, M.C., M.R. and V.D.P.; software, M.C. and A.G.; validation, M.R. and V.D.P.; formal analysis, S.M. and R.M.; investigation, M.C., M.R., V.D.P. and A.G.; resources, A.G. and R.M.; data curation, M.C.; writing—original draft preparation, M.C., M.R., V.D.P. and A.G.; writing—review and editing, S.M. and R.M.; supervision, S.M. and R.M.; project administration, S.M. and R.M. All authors have read and agreed to the published version of the manuscript.

Funding: This research received no external funding.

Data Availability Statement: Data available on request due to restrictions eg privacy or ethical.

Conflicts of Interest: The authors declare no conflict of interest.

Appendix A

Table A1. Table of errors made by the participants during the experiments.

Type of Error	Error Code	Severity Level
No errors committed during the task	E0	-
Error in the removal of panel's pop rivets	E1	Low
Error in the positioning of the tool for pop rivets' removal after usage	E2	Medium
Error in the removal of a panel's stringer	E3	Low
Error in the positioning of the panel's stringer on the work table	E4	Low
Error in the operation of deburring	E5	High
Error in the positioning of the deburring tool after usage	E6	Medium
Error in positioning the sealant application tool after usage	E7	Low
Error in releasing the stringer during the transportation to the panel	E8	Medium
Error in positioning the stringer on the panel	E9	Medium
Error during the grasp of elements supporting the stringer	E10	Low
Error in positioning the elements supporting the stringer	E11	Low

References

1. Devol, G. Mechanical Arm. U.S. Patent 2,988,237, 16 June 1961. Available online: https://patents.google.com/patent/US2988237A/en (accessed on 6 June 2023).
2. Humlum, A. *Robot Adoption and Labor Market Dynamics*; Rockwool Foundation Research Unit: Berlin, Germany, 2022.
3. Bartneck, C.; Belpaeme, T.; Eyssel, F.; Kanda, T.; Keijsers, M.; Šabanović, S. *Human-Robot Interaction: An Introduction*; Cambridge University Press: Cambridge, UK, 2020.

4. Salem, M.; Dautenhahn, K. Evaluating trust and safety in HRI: Practical issues and ethical challenges. In *Emerging Policy and Ethics of Human-Robot Interaction*; University of Hertfordshire: London, UK, 2015.
5. De Santis, A.; Siciliano, B.; De Luca, A.; Bicchi, A. An atlas of physical human–robot interaction. *Mech. Mach. Theory* **2008**, *43*, 253–270. [CrossRef]
6. Zacharaki, A.; Kostavelis, I.; Gasteratos, A.; Dokas, I. Safety bounds in human robot interaction: A survey. *Saf. Sci.* **2020**, *127*, 104667. [CrossRef]
7. Sharkawy, A.N.; Koustoumpardis, P.N. Human–robot interaction: A review and analysis on variable admittance control, safety, and perspectives. *Machines* **2022**, *10*, 591. [CrossRef]
8. Caterino, M.; Chiacchio, P.; Cristalli, C.; Fera, M.; Lettera, G.; Natale, C.; Nisi, M. Robotized assembly and inspection of composite fuselage panels: The LABOR project approach. In *IOP Conference Series: Materials Science and Engineering*; IOP Publishing: Bristol, UK, 2021; Volume 1024, p. 012019.
9. *ISO/TS 15066*; Robots and Robotic Devices: Collaborative Robots. International Organization for Standardization: Geneva, Switzerland, 2016.
10. Lee, J.D.; See, K.A. Trust in automation: Designing for appropriate reliance. *Hum. Factors* **2004**, *46*, 50–80. [CrossRef]
11. Rossi, A.; Dautenhahn, K.; Koay, K.L.; Walters, M.L. How the timing and magnitude of robot errors influence peoples' trust of robots in an emergency scenario. In Proceedings of the Social Robotics: 9th International Conference, ICSR 2017, Tsukuba, Japan, 22–24 November 2017; pp. 42–52.
12. Rossi, A.; Dautenhahn, K.; Koay, K.L.; Walters, M.L.; Holthaus, P. Evaluating people's perceptions of trust in a robot in a repeated interactions study. In Proceedings of the Social Robotics: 12th International Conference, ICSR 2020, Golden, CO, USA, 14–18 November 2020; pp. 453–465.
13. Charalambous, G.; Fletcher, S.; Webb, P. The development of a scale to evaluate trust in industrial human-robot collaboration. *Int. J. Soc. Robot.* **2016**, *8*, 193–209. [CrossRef]
14. Koppenborg, M.; Nickel, P.; Naber, B.; Lungfiel, A.; Huelke, M. Effects of movement speed and predictability in human–robot collaboration. *Hum. Factors Ergon. Manuf. Serv. Ind.* **2017**, *27*, 197–209. [CrossRef]
15. Kadir, B.A.; Brober, O. Human-centered Design of Work Systems in the Transition to Industry 4.0. *Appl. Ergon.* **2021**, *92*, 103334. [CrossRef]
16. Caterino, M.; Rinaldi, M.; Fera, M. Digital ergonomics: An evaluation framework for the ergonomic risk assessment of heterogeneous workers. *Int. J. Comput. Integr. Manuf.* **2022**, 1–21. [CrossRef]
17. Cárdenas-Robledo, L.A.; Hernández-Uribe, Ó.; Reta, C.; Cantoral-Ceballos, J.A. Extended reality applications in industry 4.0.-A systematic literature review. *Telemat. Inform.* **2022**, *73*, 101863. [CrossRef]
18. Or, C.K.; Duffy, V.G.; Cheung, C.C. Perception of safe robot idle time in virtual reality and real industrial environments. *Int. J. Ind. Ergon.* **2009**, *39*, 807–812. [CrossRef]
19. Ottogalli, K.; Rosquete, D.; Rojo, J.; Amundarain, A.; María Rodríguez, J.; Borro, D. Virtual reality simulation of human-robot coexistence for an aircraft final assembly line: Process evaluation and ergonomics assessment. *Int. J. Comput. Integr. Manuf.* **2021**, *34*, 975–995. [CrossRef]
20. Zio, E. Reliability engineering: Old problems and new challenges. *Reliab. Eng. Syst. Saf.* **2009**, *94*, 125–141. [CrossRef]
21. Swain, A.D.; Guttmann, H.E. *Handbook of Human-Reliability Analysis with Emphasis on Nuclear Power Plant Applications*; Final report (No. NUREG/CR-1278; SAND-80-0200); Sandia National Labs: Albuquerque, NM, USA, 1983.
22. Di Pasquale, V.; Miranda, S.; Iannone, R.; Riemma, S. A simulator for human error probability analysis (SHERPA). *Reliab. Eng. Syst. Saf.* **2015**, *139*, 17–32. [CrossRef]
23. Hou, L.X.; Liu, R.; Liu, H.C.; Jiang, S. Two decades on human reliability analysis: A bibliometric analysis and literature review. *Ann. Nucl. Energy* **2021**, *151*, 107969. [CrossRef]
24. Kern, C.; Refflinghaus, R. Cross-disciplinary method for predicting and reducing human error probabilities in manual assembly operations. *Total Qual. Manag. Bus. Excell.* **2013**, *24*, 847–858. [CrossRef]
25. Kern, C.; Refflinghaus, R. Assembly-specific database for predicting human reliability in assembly operations. *Total Qual. Manag. Bus. Excell.* **2015**, *26*, 1056–1070. [CrossRef]
26. Caputo, A.C.; Pelagagge, P.M.; Salini, P. Modelling human errors and quality issues in kitting processes for assembly lines feeding. *Comput. Ind. Eng.* **2017**, *111*, 492–506. [CrossRef]
27. Di Pasquale, V.; Miranda, S.; Iannone, R.; Riemma, S. An HRA-based simulation model for the optimization of the rest breaks configurations in human-intensive working activities. *IFAC-PapersOnLine* **2015**, *48*, 332–337. [CrossRef]
28. Torres, Y.; Nadeau, S.; Landau, K. Classification and quantification of human error in manufacturing: A case study in complex manual assembly. *Appl. Sci.* **2021**, *11*, 749. [CrossRef]
29. Wang, Y.B.; Zhang, T.; Xue, Q. The design and realization of HRA service system for the engine assembly. In *Applied Mechanics and Materials*; Trans Tech Publications Ltd.: Wollerau, Switzerland, 2014; Volume 635, pp. 1900–1905.
30. Pan, X.; Lin, Y.; He, C. A review of cognitive models in human reliability analysis. *Qual. Reliab. Eng. Int.* **2017**, *33*, 1299–1316. [CrossRef]
31. Mosleh, A.; Chang, Y.H. Model-based human reliability analysis: Prospects and requirements. *Reliab. Eng. Syst. Saf.* **2004**, *83*, 241–253. [CrossRef]

32. Sauppé, A.; Mutlu, B. The social impact of a robot co-worker in industrial settings. In Proceedings of the 33rd Annual ACM Conference on Human Factors in Computing Systems, Seoul, Republic of Korea, 18–23 April 2015; pp. 3613–3622.
33. Laudante, E.; Greco, A.; Caterino, M.; Fera, M. Human–robot interaction for improving fuselage assembly tasks: A case study. *Appl. Sci.* **2020**, *10*, 5757. [CrossRef]
34. Lu, L.; Xie, Z.; Wang, H.; Li, L.; Xu, X. Mental stress and safety awareness during human-robot collaboration-Review. *Appl. Ergon.* **2022**, *105*, 103832. [CrossRef] [PubMed]
35. Stanton, N.A.; Chambers, P.R.; Piggott, J. Situational awareness and safety. *Saf. Sci.* **2001**, *39*, 189–204. [CrossRef]
36. Gasparetto, A.; Zanotto, V. A technique for time-jerk optimal planning of robot trajectories. *Robot. Comput-Integr. Manuf.* **2008**, *24*, 415–426. [CrossRef]
37. Di Pasquale, V.; De Simone, V.; Giubileo, V.; Miranda, S. A taxonomy of factors influencing worker's performance in human–robot collaboration. *IET Collab. Intell. Manuf.* **2023**, *5*, e12069. [CrossRef]
38. Alves, C.; Cardoso, A.; Colim, A.; Bicho, E.; Braga, A.C.; Cunha, J.; Faria, C.; Rocha, L.A. Human–robot interaction in industrial settings: Perception of multiple participants at a crossroad intersection scenario with different courtesy cues. *Robotics* **2022**, *11*, 59. [CrossRef]
39. Bethel, C.L.; Salomon, K.; Murphy, R.R.; Burke, J.L. Survey of psychophysiology measurements applied to human-robot interaction. In Proceedings of the RO-MAN 2007-The 16th IEEE International Symposium on Robot and Human Interactive Communication, Jeju, Republic of Korea, 26–29 August 2007; pp. 732–737.
40. Huber, M.; Rickert, M.; Knoll, A.; Brandt, T.; Glasauer, S. Human-robot interaction in handing-over tasks. In Proceedings of the RO-MAN 2008-The 17th IEEE International Symposium on Robot and Human Interactive Communication, Munich, Germany, 1–3 August 2008; pp. 107–112.
41. Zhang, B.; Amirian, J.; Eberle, H.; Pettré, J.; Holloway, C.; Carlson, T. From HRI to CRI: Crowd Robot Interaction—Understanding the Effect of Robots on Crowd Motion: Empirical Study of Pedestrian Dynamics with a Wheelchair and a Pepper Robot. *Int. J. Soc. Robot.* **2021**, *14*, 631–643. [CrossRef]
42. Chen, Z.; Jiang, C.; Guo, Y. Pedestrian-robot interaction experiments in an exit corridor. In Proceedings of the 2018 15th International Conference on Ubiquitous Robots (UR), Honolulu, HI, USA, 26–30 June 2018; pp. 29–34.
43. Xie, Y.; Bodala, I.P.; Ong, D.C.; Hsu, D.; Soh, H. Robot capability and intention in trust-based decisions across tasks. In Proceedings of the 2019 14th ACM/IEEE International Conference on Human-Robot Interaction (HRI), Daegu, Republic of Korea, 11–14 March 2019; pp. 39–47.
44. Vassallo, C.; Olivier, A.H.; Souères, P.; Crétual, A.; Stasse, O.; Pettré, J. How do walkers avoid a mobile robot crossing their way? *Gait Posture* **2017**, *51*, 97–103. [CrossRef]
45. De Simone, V.; Di Pasquale, V.; Giubileo, V.; Miranda, S. Human-Robot Collaboration: An analysis of worker's performance. *Procedia Comput. Sci.* **2022**, *200*, 1540–1549. [CrossRef]
46. Hashemi-Petroodi, S.E.; Thevenin, S.; Kovalev, S.; Dolgui, A. Operations management issues in design and control of hybrid human-robot collaborative manufacturing systems: A survey. *Annu. Rev. Control* **2020**, *49*, 264–276. [CrossRef]
47. Morioka, M.; Sakakibara, S. A new cell production assembly system with human–robot cooperation. *CIRP Ann.* **2010**, *59*, 9–12. [CrossRef]
48. Weems, C.F.; Costa, N.M.; Watts, S.E.; Taylor, L.K.; Cannon, M.F. Cognitive errors, anxiety sensitivity, and anxiety control beliefs: Their unique and specific associations with childhood anxiety symptoms. *Behav. Modif.* **2007**, *31*, 174–201. [CrossRef] [PubMed]
49. Barker, L.M.; Nussbaum, M.A. The effects of fatigue on performance in simulated nursing work. *Ergonomics* **2011**, *54*, 815–829. [CrossRef]
50. Rahimi, M.; Karwowski, W. Human perception of robot safe speed and idle time. *Behav. Inf. Technol.* **1990**, *9*, 381–389. [CrossRef]
51. Dehais, F.; Sisbot, E.A.; Alami, R.; Causse, M. Physiological and subjective evaluation of a human–robot object hand-over task. *Appl. Ergon.* **2011**, *42*, 785–791. [CrossRef]
52. Taherdoost, H. What Is the Best Response Scale for Survey and Questionnaire Design; Review of Different Lengths of Rating Scale/Attitude Scale/Likert Scale. *Int. J. Acad. Res. Manag.* **2019**, *8*, 1–10.
53. De Barros RS, M.; Hidalgo JI, G.; de Lima Cabral, D.R. Wilcoxon rank sum test drift detector. *Neurocomputing* **2018**, *275*, 1954–1963. [CrossRef]

machines

MDPI

Article

Experimental Assessment of Hole Quality and Tool Condition in the Machining of an Aerospace Alloy

Muhammad Aamir [1,*], Aamer Sharif [2], Muhammad Zeeshan Zahir [3], Khaled Giasin [4] and Majid Tolouei-Rad [1]

1 School of Engineering, Edith Cowan University, Joondalup, WA 6027, Australia; m.rad@ecu.edu.au
2 Department of Mechanical Engineering, CECOS University of Information Technology and Emerging Sciences, Peshawar 25000, Pakistan; aamer@cecos.edu.pk
3 Department of Mechanical Engineering, University of Engineering and Technology, Peshawar 25000, Pakistan; zeeshan.zahir@uetpeshawar.edu.pk
4 School of Mechanical and Design Engineering, University of Portsmouth, Portsmouth PO1 3DJ, UK; khaled.giasin@port.ac.uk
* Correspondence: m.aamir@ecu.edu.au

Abstract: This paper deals with an experimental investigation of hole quality in Al2024-T3, which is one of the aerospace alloys used in aircraft fuselage skin due to its high level of resistance to fatigue crack propagation. The experiments are conducted with 6 mm uncoated carbide and HSS drill bits using a CNC machine under dry conditions and different drilling parameters. The characteristics of the hole quality are investigated in terms of its perpendicularity, cylindricity, circularity and hole size. An ANOVA (analysis of variance) and Pareto charts are used to analyze the effects of the drilling parameters on the hole quality. The hole quality is also assessed using a digital microscope to observe the formation of hole burrs. Moreover, scanning electron microscopy is also used to investigate the inside-hole surface defects. Further investigations are carried out using optical microscopy to inspect the post-drilling tool condition at high drilling parameters. The results show that hole quality reduces as the feed rate and spindle speed increase. However, from the ANOVA results and Pareto charts, the influence of the feed rate on the hole quality is found to be insignificant. At the same time, the type of drill bit material shows the highest percentage of contribution affecting the hole quality, following the spindle speed. The HSS drill bit shows more adhesion and built-up edges than the uncoated carbide drill bit. There were more burrs formed at the hole edges when the holes were drilled with uncoated HSS drill bits. In the same way, the SEM analysis reveals more surface deformation and damage defects inside the hole walls of holes drilled using the uncoated HSS drill bit.

Keywords: drilling; Al2024-T3; hole quality; ANOVA; tool condition

1. Introduction

There is no doubt about how important it is to perform drilling operations in various industries to make a profit and survive in today's very competitive market because most industrial products in our daily lives incorporate holes generated via drilling operations. Figure 1 shows the importance of drilling in all machining processes in various industries [1]. However, the drilling process becomes challenging when many holes are required, like in the aerospace industry [2]. For instance, in aircraft bodies, about 80% of fatigue cracks are because of poor connecting holes, and 50–90% of fractures in aging planes are due to fatigue fractures of fastener holes [3]. Hence, to ensure high-precision structural integrity in aerospace alloys, the quality of holes drilled in aluminum is significantly important [4,5]. Furthermore, the major problems in the drilling process include high cutting forces, low hole quality, tool wear, etc., which require selecting the appropriate tool, increasing the cost for the manufacturing sectors [6,7]. The built-up edge is also one of the problems associated with dry drilling that ultimately affects the tool and thus reduces the hole quality; however, drilling in dry conditions is environmentally friendly as it reduces the environmental effect of coolants [8,9]. Hence, to achieve a long tool life and high dimensional accuracy, drilling performance is therefore dependent on a variety of variables, including tool geometry, tool materials, drilling parameters such as the spindle speed *(n)* and feed rate *(f)*, the type of drilling machine, and the absence or presence of coolants [10,11]. Therefore, researchers are interested in optimizing/investigating these important drilling parameters to maximize productivity.

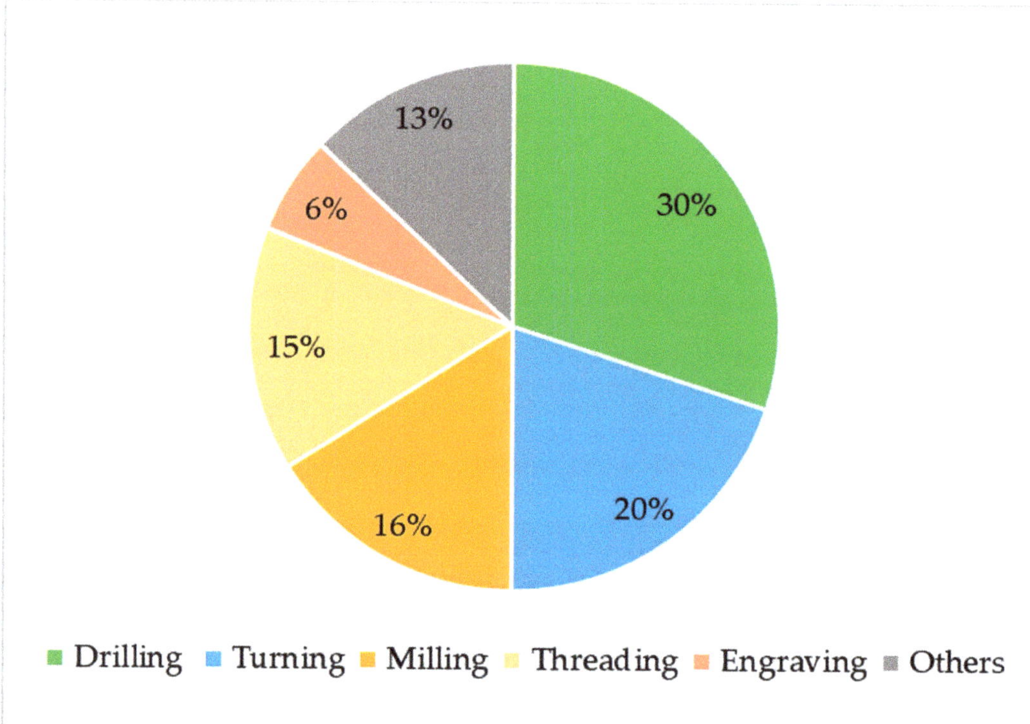

Figure 1. Machining processes.

Previously, Dahnel et al. [12] performed drilling experiments on the tool wear on Al7075 using tungsten carbide tools with n values of 4000, 6000, and 8000 rpm and f values from 0.01 to 0.10 mm/rev. It was concluded in their study that a lower spindle speed could reduce heat generation; hence, the lower cutting speed was recommended for dry drilling Al7075. Islam and Boswell [13] used high-speed steel (HSS) tools to determine the effect of drilling parameters and cooling methods on the quality of holes drilled in Al6061. The investigation was carried out by measuring the surface roughness, diameter error, and circularity. The cooling methods included cryogenic drilling, flood drilling, and minimum-quality lubrication (MQL). The study concluded that diameter error was highly affected by the cooling method, followed by the surface roughness and circularity. MQL drilling produced good surface roughness and dimensional accuracy. In contrast, the circularity was best achieved using cryogenic drilling. Khunt et al. [14] assessed the drilling performance of AA6063. The experiments were performed using HSS drill bits. Different cutting environments, including flood, MQL-sunflower, dry, flood cooling, and MQL-castor oil, were selected to measure torque, surface roughness, and axial thrust. The results concluded that surface roughness was improved with vegetable MQL, while the torque and axial thrust were reduced. Luo et al. [15] established a finite element model using different spindle and feed speeds to drill Al7075-T6. They concluded that an increase in the feed speed resulted in high axial force and torque. Similarly, the thickness of the uncut chips also increased. Furthermore, the numerical simulation showed a higher tool temperature field at the main cutting edge than at the cross edge. Pramanik et al. [16] worked on drilling Al6061-T6 in three different conditions, including liquid nitrogen (LN2), compressed air and MQL, to identify their impact on chip formation, the quality of holes, active peak power, and surface roughness. The study concluded that cooling techniques had a negligible impact on active power consumption. Additionally, the coolant's impact on the chip thickness ratio and surface roughness was unclear. However, the surface roughness increased as the speeds and feeds increased, while the chip thickness ratio increased with speed and decreased with the feed. In a study by Banerjee et al. [17], 102° and 115° point angles were recommended for the minimum burr height and thickness when drilling aluminum using high-speed steel (HSS) drill bits. Regarding the previous studies on Al2024, the selection of various drilling parameters is provided in Table 1.

Table 1. Previous studies on drilling Al2024.

Material	Spindle Speed/ Cutting Speed	Feed Rate	Drill Bits	Areas Studied	Ref.
Al2024	1000, 2000, and 3000 (rpm)	0.04, 0.08, and 0.14 (mm/rev)	Uncoated carbide, 6 mm and 10 mm	Thrust force, surface roughness, burrs, hole surface damage analysis, chips formation, and tool condition	[18]
AA2024	60, 120, and 180 (m/min)	0.05, 0.15, and 0.25 (mm/rev)	HSS twist drill with cobalt, 6 mm	Surface roughness, thrust force, hole diameter, and exit burr height	[19]
Al2024	1000, 3000, 6000, and 9000 (rpm)	100, 300, 600, 900 (mm/min)	TiAlN-coated carbide twist drill	Chip formation, surface roughness, hole size, burrs, and circularity error	[20]
Al2024	28 and 94 (m/min)	0.04 (mm/rev)	HSS and HSS-Co	Thrust force, torque, and surface finish	[21]
Al 2024	30, 45, and 60 (m/min)	0.15, 0.20, and 0.25 (mm/rev)	Uncoated HSS, TiN and TiAlN-coated	Surface finish and hole diameter	[22]

Hence, studies in the literature are limited to other drilling process parameters, cooling techniques, or materials. There is also a lack of studies investigating the significant characteristics of hole quality. Therefore, this study evaluates important hole characteristics, including circularity, hole size, perpendicularity, and cylindricity. The analysis was carried out using different types of drill bit materials, spindle speeds, and feed rates. The results were then evaluated using Pareto charts and an analysis of variance (ANOVA). The study also included further experiments to examine the post-drilling tool conditions at high drilling parameters and the burrs around the hole edges using optical microscopy. A further examination of the quality of holes was performed using scanning electron microscopy to investigate any defects inside the drilled hole surfaces.

2. Materials and Methods

Al2024-T3 and a CNC machine were used for drilling experiments. CNC machines are high-volume production machines used for achieving high levels of productivity without compromising quality. Uncoated HSS and uncoated carbide drill bits with a size of 6 mm were mounted on the CNC machine to drill holes in the Al2024-T3. The drilling experiments were repeated three times, and each time, a new drill bit was used for the accuracy of the results. Hence, a total of 27 holes were drilled under dry conditions. The workpiece material, drilling parameters, and experimental details are provided in Table 2.

Table 2. Experimental details.

Material Details									
Material	Al2024-T3								
Dimension	150 × 200 mm^2								
Thickness	10 mm								
Chemical composition	Mg	Cr	Si	Z	Cu	Mn	Fe	Ti	Al
	1.5	0.1	0.5	0.25	4.5	0.6	0.5	0.15	Balance
Ultimate tensile strength	445 MPa								
Drilling parameters									
Feed rate (mm/rev)	0.04, 0.08, 0.14								
Spindle speed (rpm)	1500, 2500, 3500								
Drilling condition	Dry								
Drill bit details									
Type	Twist drill								
Material	HSS, carbide								
Coating	None								
Drill diameter	6 mm								
Shank diameter	6 mm								
Helix angle	30°								
Number of flutes	2 mm								
Machines used									
Machine tool	CNC								
Hole size, circularity, cylindricity, and perpendicularity	Coordinate measuring machine (Taichung, Taiwan)								
Burrs	USB digital microscope								
Hole surface defects	Scanning electron microscopy (Hitachi SU5000 Chiyoda, Japan)								
Tool condition	Optical microscope (LEICA M80)								

The HSS and carbide drills are the most frequently used drill bits in industries, based on the application. A likely explanation for selecting HSS drill bits is that they are less expensive, commonly used, robust, and durable. They are also tough and heat-resistant. In contrast, carbide drills are extremely hard, with a greater heat dissipation rate, and are considered the toughest compared to other drill bits [23]. In addition, the 6 mm size is a commonly used tool size range for aerospace alloys. At the same time, a high helix angle was previously suggested to achieve high-quality holes during the drilling of Al20254-T3 [24]. Furthermore, according to the machinery handbook [25], the feed rate should be 0.05–0.15 mm/rev when a twist drill with a size of 3.175–6.35 mm is used. Finally, Al2024-T3 was selected because it is a highly used aluminum alloy in the aerospace industry, where millions of holes are required. Hence, the drilling process and geometric tolerances are of great interest [26].

A coordinate measuring machine (CMM) was used to measure perpendicularity, circularity, cylindricity, and hole size to investigate hole quality. It is worth noting that CMMs are mostly used in the automotive, aerospace and defense industries to measure geometric features of parts [27]. Afterwards, at a confidence interval of 95%, the impacts of the drilling parameters on the hole quality characteristics were evaluated using an ANOVA.

The study also included additional experiments to examine the post-drilling tool conditions for each drill bit at high spindle speed of 3500 rpm and feed rate of 0.14 mm/rev. The investigation was carried out using optical microscopy after 9, 27, and 45 holes, and a comparison was made between the built-up edges generated on each drill bit. Afterwards, scanning electron microscopy (SEM) was used to examine the holes. The study also included the examination of the burrs around the holes using a digital microscope. Further examinations of the top and bottom hole edges and the inside surface defects of the holes were performed via SEM.

3. Experimental Results and Analysis of Variance

3.1. Hole Deviation and Circularity

Figures 2 and 3 show the average hole size and circularity using the uncoated HSS and uncoated carbide drill bits with different spindle speeds and feed rates. In general, the results show that hole deviation and circularity increase as the spindle speed and feed rate increase. At the same time, the uncoated carbide drills produced holes with lesser deviations from the hole's nominal size and fewer errors in circularity. The hole deviation from the nominal size (6 mm) for the uncoated HSS drill bit ranged from 6.0581 mm to 6.0753 mm, while for the uncoated carbide drill bits, the minimum value obtained was 6.0086 mm, while the maximum was 6.0281 mm. Similarly, the circularity ranged from 0.0261 to 0.0541 mm for the uncoated HSS drill bits. In contrast, it ranged from 0.0193 to 0.0284 mm when the carbide drill bits were used. In the current study, the ANOVA was used at a 95% confidence interval, so only those values with a p-value of less than 0.05 were considered significant. Therefore, the results in Table 3 indicate that the greatest impact on hole size and circularity was achieved by the drill bit materials. The drill bit's material type affected the hole size by 94.10% and the circularity by 45.07%. The spindle speed's influence on the hole size was 3.75%, while none of the other parameters affected the hole size and circularity in the 95% confidence interval.

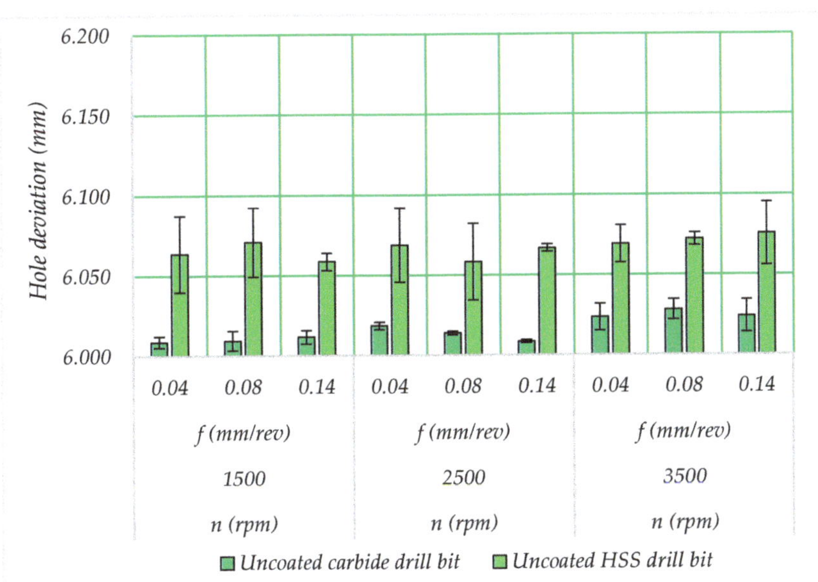

Figure 2. Hole size.

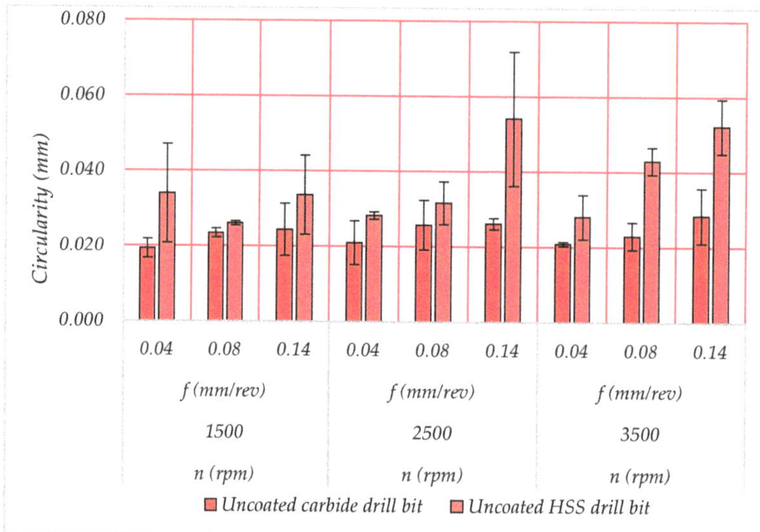

Figure 3. Circularity.

Table 3. ANOVA for hole size and circularity.

Hole Size							
Source	DF	Seq SS	Adj SS	Adj MS	F-Value	p-Value	Contribution
Model	13	0.012127	0.012127	0.000933	33.73	0.002	99.10%
Linear	5	0.01198	0.01198	0.002396	86.63	0	97.89%
Drill bit material	1	0.011516	0.011516	0.011516	416.38	0	94.10%
Spindle speed	2	0.000459	0.000459	0.000229	8.29	0.038	3.75%
Feed rate	2	0.000006	0.000006	0.000003	0.11	0.9	0.05%
Two-way interactions	8	0.000147	0.000147	0.000018	0.66	0.712	1.20%
Drill bit material × spindle speed	2	0.000041	0.000041	0.000021	0.74	0.532	0.34%
Drill bit material × feed rate	2	0.000005	0.000005	0.000002	0.08	0.922	0.04%
Spindle speed × feed rate	4	0.000101	0.000101	0.000025	0.92	0.533	0.83%
Error	4	0.000111	0.000111	0.000028	-	-	0.90%
Total	17	0.012238	-	-	-	-	100.00%
Circularity							
Model	13	0.001579	0.001579	0.000121	3.24	0.133	91.33%
Linear	5	0.001282	0.001282	0.000256	6.84	0.043	74.19%
Drill bit material	1	0.000779	0.000779	0.000779	20.79	0.01	45.07%
Spindle speed	2	0.000107	0.000107	0.000054	1.43	0.34	6.21%
Feed rate	2	0.000396	0.000396	0.000198	5.28	0.075	22.91%
Two-way interactions	8	0.000296	0.000296	0.000037	0.99	0.544	17.14%
Drill bit material × spindle speed	2	0.00005	0.00005	0.000025	0.67	0.563	2.89%
Drill bit material × feed rate	2	0.000115	0.000115	0.000057	1.53	0.321	6.63%
Spindle speed × feed rate	4	0.000132	0.000132	0.000033	0.88	0.548	7.63%
Error	4	0.00015	0.00015	0.000037	-	-	8.67%
Total	17	0.001728	-	-	-	-	100.00%

3.2. Cylindricity and Perpendicularity

In the current study, the cylindricity and perpendicularity increased with increases in the drilling parameters. However, a significant impact on both cylindricity and perpendicularity was found based on the type of drill bit material used, as shown in Figures 4 and 5. The uncoated HSS drill bit holes have greater cylindricity and perpendicularity than the uncoated carbide drill bit holes. Hence, when using the carbide drill bits, the lowest cylindricity and perpendicularity values produced were 0.0271 mm and 0.0120 mm. Likewise, the ANOVA result in Table 4 indicates that the impact of drill bit material on the cylindricity was 64.35%, while the impact on perpendicularity was 61.08%. Furthermore, the spindle speed had an impact of 16.17% on cylindricity and 8.92% on perpendicularity. However, the feed rate showed an insignificant contribution in affecting the cylindricity and perpendicularity in this study at the confidence interval of 95%. Moreover, in the two-way interactions, the combination of drill type and spindle speed had an influence of 5.32% on the cylindricity. In comparison, the combined effect of drill type and feed rate had an impact of 5.78% on the cylindricity. At the same time, in the two-way interactions, only the drill type and spindle affected perpendicularity, with a 13.06% contribution. Hence, it is concluded that the drill bit material plays a significant role during the drilling process in affecting the hole quality characteristics.

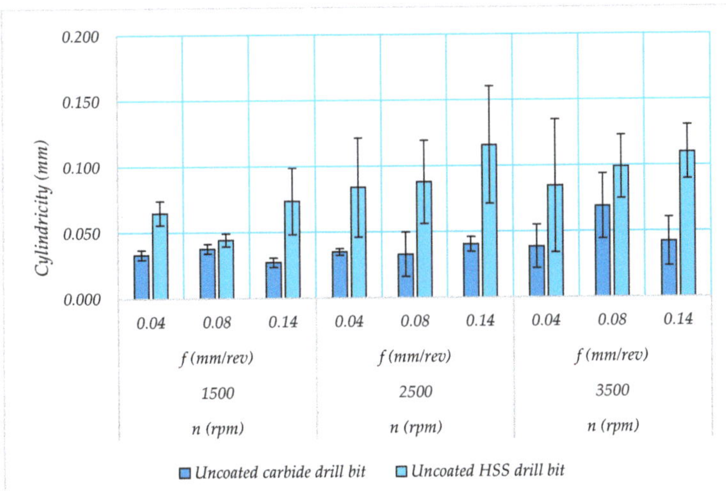

Figure 4. Cylindricity.

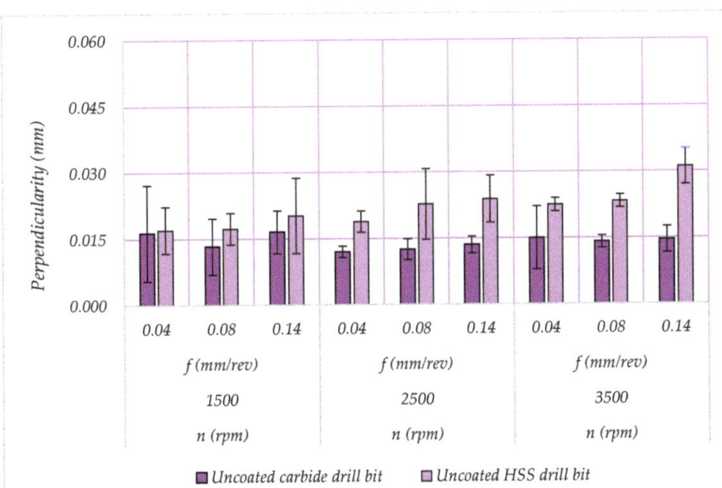

Figure 5. Perpendicularity.

Table 4. ANOVA for cylindricity and perpendicularity.

Cylindricity							
Source	DF	Seq SS	Adj SS	Adj MS	F-Value	*p*-Value	Contribution
Model	13	0.014233	0.014233	0.001095	32.18	0.002	99.05%
Linear	5	0.011966	0.011966	0.002393	70.34	0.001	83.28%
Drill bit material	1	0.009246	0.009246	0.009246	271.76	0	64.35%
Spindle speed	2	0.002323	0.002323	0.001162	34.14	0.003	16.17%
Feed rate	2	0.000396	0.000396	0.000198	5.82	0.065	2.76%
Two-Way Interactions	8	0.002267	0.002267	0.000283	8.33	0.029	15.78%
Drill bit material × spindle speed	2	0.000765	0.000765	0.000382	11.24	0.023	5.32%
Drill bit material × feed rate	2	0.000831	0.000831	0.000415	12.21	0.02	5.78%
Spindle speed × feed rate	4	0.000671	0.000671	0.000168	4.93	0.076	4.67%
Error	4	0.000136	0.000136	0.000034	-	-	0.95%
Total	17	0.014369	-	-	-	-	100.00%
Perpendicularity							
Model	13	0.00042	0.00042	0.000032	12.89	0.012	97.67%
Linear	5	0.000334	0.000334	0.000067	26.69	0.004	77.74%
Drill bit material	1	0.000262	0.000262	0.000262	104.83	0.001	61.08%
Spindle speed	2	0.000038	0.000038	0.000019	7.65	0.043	8.92%
Feed rate	2	0.000033	0.000033	0.000017	6.65	0.053	7.75%
Two-Way interactions	8	0.000086	0.000086	0.000011	4.27	0.088	19.93%
Drill bit material × spindle speed	2	0.000056	0.000056	0.000028	11.21	0.023	13.06%
Drill bit material × feed rate	2	0.00002	0.00002	0.00001	3.97	0.112	4.63%
Spindle speed × feed rate	4	0.00001	0.00001	0.000002	0.96	0.516	2.23%
Error	4	0.00001	0.00001	0.000003	-	-	2.33%
Total	17	0.00043	-	-	-	-	100.00%

4. Discussion

In the current study, the quality of hole characteristics reduces with increases in the spindle speed and feed rate. However, at a 95% confidence interval, the ANOVA results concluded that the most influencing parameter affecting the hole quality was the type of drill bit material, followed by the spindle speed. In contrast, the influence of the feed rate on the quality of the holes was found to be insignificant. This could also be confirmed by the Pareto chart, as provided in Figure 6. Additionally, the uncoated carbide drill bit showed low perpendicularity, low circularity, low cylindricity, and a low level of deviation of the hole from the nominal size compared to the results produced by the uncoated HSS drill bits. One of the reasons for this might be the formation of fewer burrs around the entry and exit sides of the holes produced by the uncoated carbide drill bits. Thus, the burrs around the holes were observed using an optical microscope, as shown in Figure 7a, which is an additional method apart from using the profile meters [24]. Hence, the formation of burrs is provided in Figure 7b. Further details can be found in Appendix A. It is also worth noting that in this study, the burrs at the edges of the holes were small and did require further processing, such as de-burring. However, the burrs can affect the geometric tolerances and might be affected by the drilling parameters. Hence, the material was further examined using scanning electron microscopy at the entry and exit sides of the hole edges, as shown in Figure 8. The SEM examination was also extended to the inside hole surface, which revealed that the surface deformation, damage defects, and chip adhesion were more visible in the holes drilled with uncoated HSS drill bits, as provided in Figure 9. It is also worth mentioning that previously [24], the uncoated carbide drill bits also performed better than the uncoated HSS drill bits in the process of drilling Al2023-T3. However, that study was based on a multi-hole drilling approach to investigate thrust force, burr formation, surface roughness, chips analysis, and tool conditions.

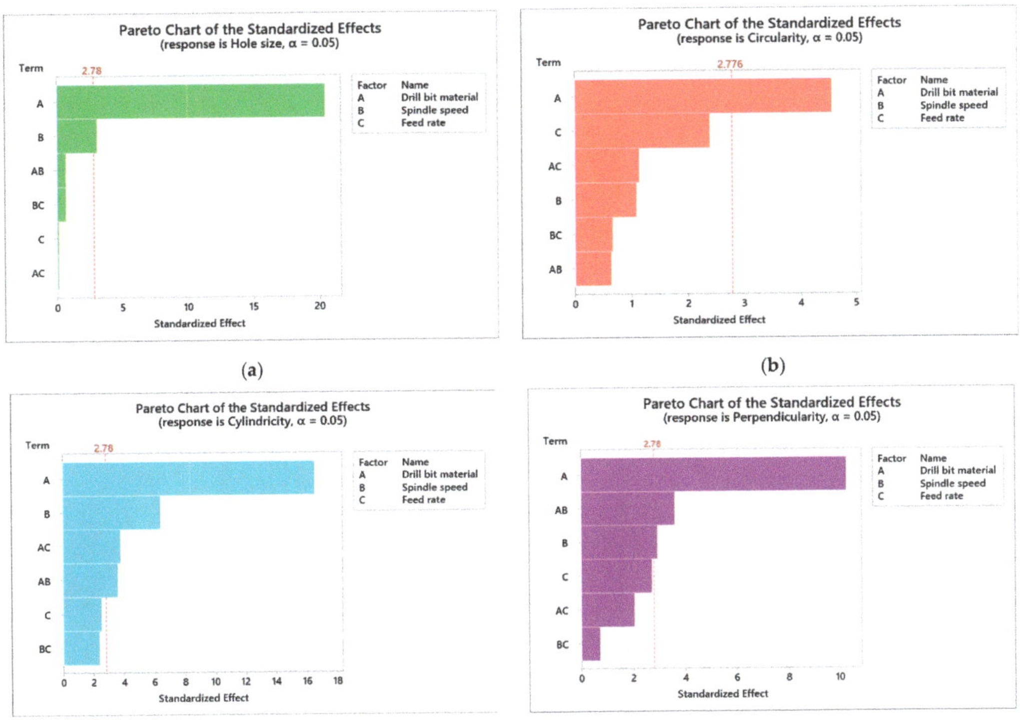

(a) (b)

(c) (d)

Figure 6. (**a**). Pareto chart for hole size. (**b**). Pareto chart for circularity. (**c**). Pareto chart for cylindricity. (**d**). Pareto chart for perpendicularity.

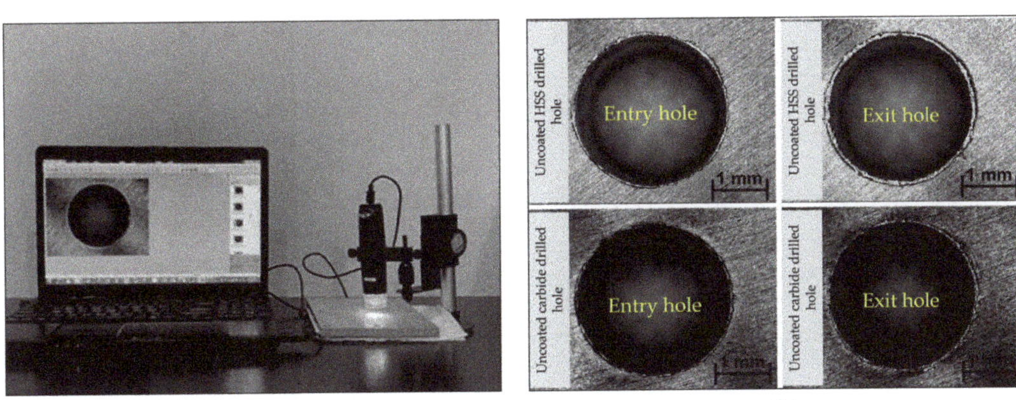

(a) (b)

Figure 7. (**a**) Burr investigation using digital microscope. (**b**) Burr formation at 3500 rpm and 0.14 mm/rev.

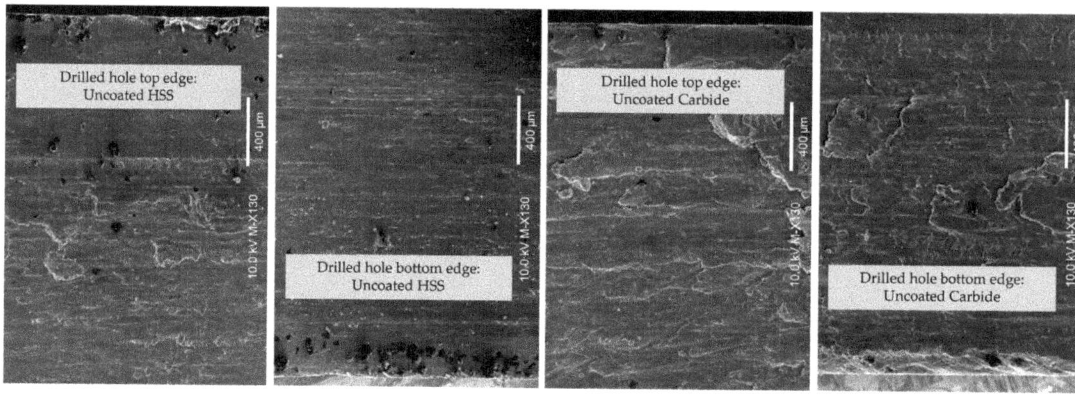

Figure 8. SEM images of hole top and bottom edge after drilling at 3500 rpm and 0.14 mm/rev.

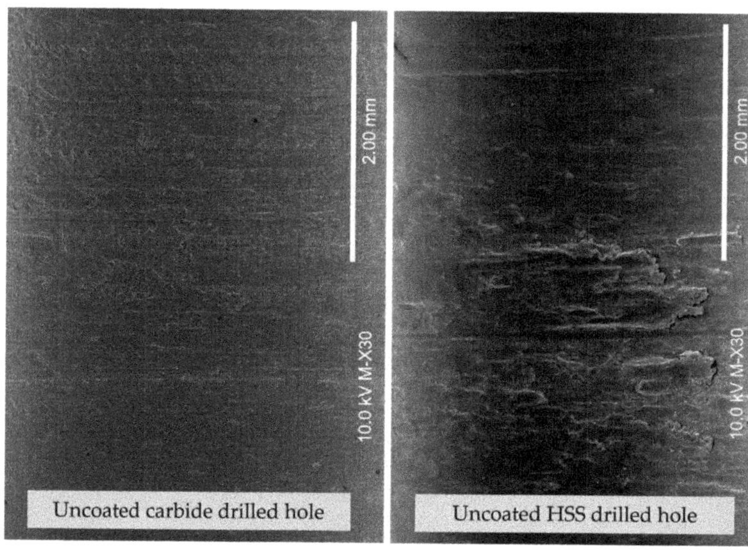

Figure 9. SEM hole analysis after drilling at 3500 rpm and 0.14 mm/rev.

Another reason for the better hole quality was the formation of a lower built-up edge (BUE) on the holes drilled with carbide drill bits. This was mainly due to the greater strength, higher hardness, and lower coefficient of friction of the carbide drill bits when compared to the uncoated HSS drill bits [24]. Furthermore, it has also been reported that high machining parameters tend to increase the temperature at the drilling zone because of the high friction and material removal rate. Hence, the material softens and undergoes plastic deformation, which causes the material to accumulate at the cutting tool, resulting in adhesion and built-up edges [28]. Previous studies have reported that high machining parameters also contribute to increasing the tool wear, deformation effects, and the quality of holes [29]. In addition, the Pareto chart in Figure 6 indicated that the second influencing factor on the hole quality was the spindle speed in this study. Therefore, additional drilling experiments were performed to compare the post-drilling tool conditions with respect to the built-up edges. Hence, a high spindle speed of 3500 rpm and a high feed rate of 0.14 mm/rev in the current drilling parameters were selected, and post-drilling conditions were investigated after 9, 27, and 45 holes were drilled. Figure 10 shows the optical microscopy images taken using each drill bit, which shows a high BUE generated on the uncoated HSS drills. It is worth mentioning that as the number of holes increases, there might be high tool work friction, resulting in a high level of energy consumption and generating higher thrust force [30]. Consequently, there are more chances of affecting the hole dimensions due to a high BUE. Hence, the SEM mid-hole surface analysis was carried out after 9, 27 and 45 holes were drilled with uncoated HSS and carbide drills, as shown in Figure 11. It is also noteworthy that the hole quality might be affected by other parameters like vibration, machine dynamics, the drilling temperature, etc. [27], which were not the scope of this study.

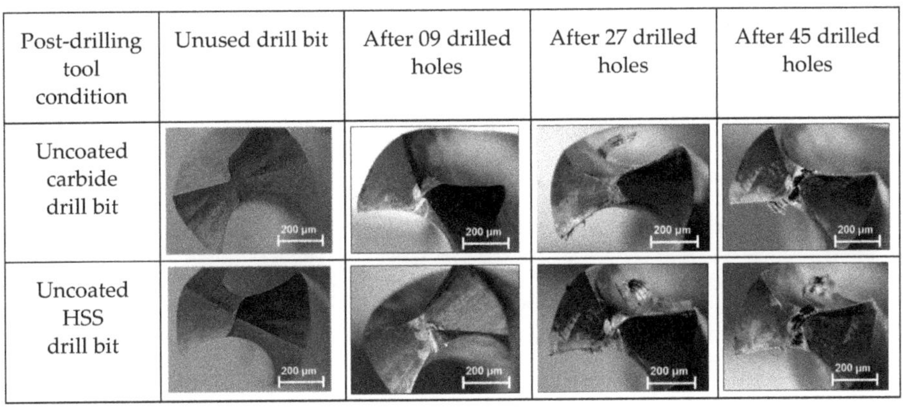

Figure 10. Post-drilling tool conditions after drilling at 3500 rpm and 0.14 mm/rev.

Figure 11. SEM mid-hole analyses after drilling at 3500 rpm and 0.14 mm/rev.

5. Conclusions

In the current study, Al2024-T3 was used for drilling experiments, and a comparison was made between uncoated HSS and carbide drill bits used with varying drilling process parameters. The investigations were based on the hole quality; however, the study was further extended to examine the tool conditions at high drilling parameters in combination with SEM analysis of the holes at the different locations. Hence, the results obtained in this study can help the scientific community and industry in selecting high-quality drilled holes. The following conclusions were made based on the above investigations.

The quality of holes with respect to hole size, cylindricity, circularity, and perpendicularity increase as the spindle speed and feed rate increase. However, ANOVA results indicated that the type of drill bit material showed the highest influence on the hole quality. The spindle speed was the second-most influencing factor affecting the hole quality. In contrast, the feed rate showed an insignificant impact on the hole quality at the confidence interval of 95% for the selected drilling parameters.

The burrs formed around the edges of holes produced by the uncoated carbide drill bits at high drilling parameters were found to be fewer than those produced by the HSS drill bits. Furthermore, the SEM images showed less surface deformation and damage defects on the hole walls drilled by the uncoated carbide drill bits. The experimental results also concluded that the uncoated carbide drill bits produced more high-quality holes than the uncoated HSS drill bits because of their high strength and resistance to wear. Similarly, the adhesion and built-up edges on the uncoated carbide drill bits were less than the uncoated HSS drill bits at high drilling parameters. Therefore, it is concluded that the drill bit materials play a significant role during the process of drilling Al2024-T3 in affecting the hole quality characteristics. Hence, drill bits with excellent wear resistance and a high degree of hardness are recommended for improving tool life.

Author Contributions: Conceptualization, M.A., M.T.-R. and K.G.; methodology, M.A. and A.S.; validation, M.A., M.T.-R. and K.G.; investigation, M.A. and K.G.; writing—original draft preparation, M.A., M.Z.Z. and A.S.; writing—review and editing, M.A., M.Z.Z. and K.G. All authors have read and agreed to the published version of the manuscript.

Funding: This research received no external funding.

Institutional Review Board Statement: Not applicable.

Informed Consent Statement: Not applicable.

Data Availability Statement: The data presented in this study are available on request.

Conflicts of Interest: The authors declare no conflict of interest.

Appendix A. Entry and Exit Hole Burrs at Varying Drilling Parameters

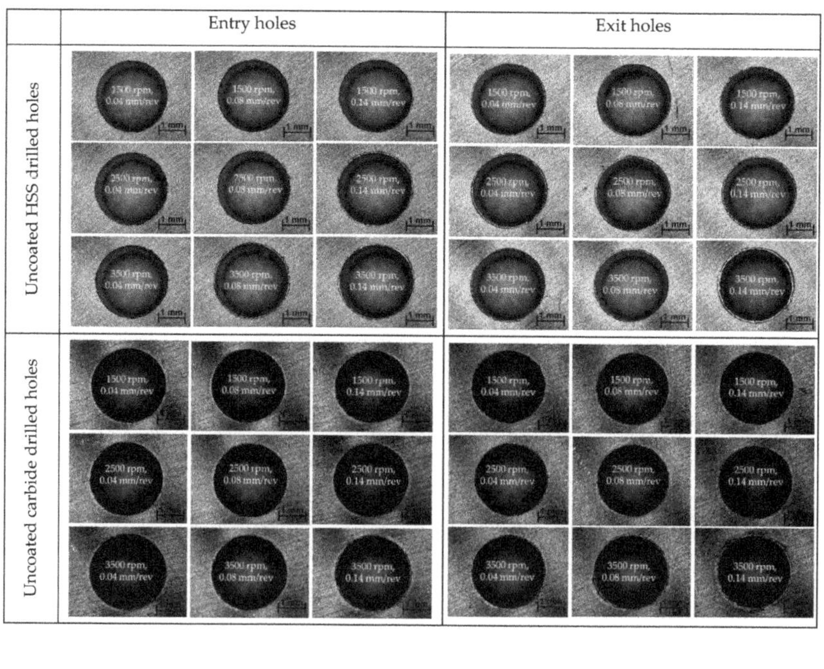

References

1. Rivero, A.; Aramendi, G.; Herranz, S.; de Lacalle, L.L. An experimental investigation of the effect of coatings and cutting parameters on the dry drilling performance of aluminium alloys. *Int. J. Adv. Manuf. Technol.* **2006**, *28*, 1–11. [CrossRef]
2. Xu, J.; Kolesnyk, V.; Li, C.; Lysenko, B.; Peterka, J.; Gupta, M.K. A critical review addressing conventional twist drilling mechanisms and quality of CFRP/Ti stacks. *J. Mater. Res. Technol.* **2023**, *24*, 6614–6651. [CrossRef]
3. Yuan, P.; Lai, T.; Li, Y.; Han, W.; Lin, M.; Zhu, Q.; Liu, Y.; Shi, Z. The attitude adjustment algorithm in drilling end-effector for aviation. *Adv. Mech. Eng.* **2016**, *8*, 1687814016629348. [CrossRef]
4. Adin, M.Ş. Performances of cryo-treated and untreated cutting tools in machining of AA7075 aerospace aluminium alloy. *Eur. Mech. Sci.* **2023**, *7*, 70–81. [CrossRef]
5. Selvakumar, S.J.; Muralidharan, S.M. Performance analysis of drills with structured surfaces when drilling CFRP/AA7075 stack under MQL condition. *J. Manuf. Process.* **2023**, *89*, 194–219. [CrossRef]
6. Habib, N.; Sharif, A.; Hussain, A.; Aamir, M.; Giasin, K.; Pimenov, D.Y.; Ali, U. Analysis of Hole Quality and Chips Formation in the Dry Drilling Process of Al7075-T6. *Metals* **2021**, *11*, 891. [CrossRef]
7. Habib, N.; Sharif, A.; Hussain, A.; Aamir, M.; Giasin, K.; Pimenov, D.Y. Assessment of Hole Quality, Thermal Analysis, and Chip Formation during Dry Drilling Process of Gray Cast Iron ASTM A48. *Eng* **2022**, *3*, 301–310. [CrossRef]
8. Dahnel, A.N.; Fauzi, M.H.; Raof, N.A.; Mokhtar, S.; Khairussaleh, N.K.M. Tool wear and burr formation during drilling of aluminum alloy 7075 in dry and with cutting fluid. *Mater. Today Proc.* **2022**, *59*, 808–813. [CrossRef]
9. Gupta, M.K.; Niesłony, P.; Korkmaz, M.E.; Kuntoğlu, M.; Królczyk, G.; Günay, M.; Sarikaya, M. Comparison of Tool Wear, Surface Morphology, Specific Cutting Energy and Cutting Temperature in Machining of Titanium Alloys under Hybrid and Green Cooling Strategies. *Int. J. Precis. Eng. Manuf.-Green Technol.* **2023**, 1–14. [CrossRef]
10. Ghasemian Fard, M.; Baseri, H.; Zolfaghari, A. Experimental investigation of the effective parameters and performance of different types of structural tools in drilling CFRP/AL alloy stacks. *J. Compos. Mater.* **2022**, *56*, 4461–4471. [CrossRef]
11. Soori, M.; Arezoo, B. Cutting tool wear minimization in drilling operations of titanium alloy Ti-6Al-4V. *Proc. Inst. Mech. Eng. Part J J. Eng. Tribol.* **2023**, *237*, 1250–1263. [CrossRef]
12. Dahnel, A.N.B.; Raof, N.A.; MK, N.K.; Mokhtar, S. The effect of cutting parameters on tool wear in drilling aluminium 7075. *Test Eng. Manag.* **2020**, *83*, 1274–1280.
13. Islam, M.N.; Boswell, B. Effect of cooling methods on hole quality in drilling of aluminium 6061-6T. *IOP Conf. Ser. Mater. Sci. Eng.* **2016**, *114*, 012022. [CrossRef]
14. Khunt, C.P.; Makhesana, M.A.; Patel, K.M.; Mawandiya, B.K. Performance assessment of vegetable oil-based minimum quantity lubrication (MQL) in drilling. *Mater. Today Proc.* **2021**, *44*, 341–345. [CrossRef]
15. Luo, H.; Fu, J.; Wu, T.; Chen, N.; Li, H. Numerical Simulation and Experimental Study on the Drilling Process of 7075-t6 Aerospace Aluminum Alloy. *Materials* **2021**, *14*, 553. [CrossRef] [PubMed]
16. Pramanik, A.; Basak, A.K.; Prakash, C.; Shankar, S.; Chattopadhyaya, S. Sustainability in drilling of aluminum alloy. *Clean. Mater.* **2022**, *3*, 100048. [CrossRef]
17. Banerjee, B.; Dhar, A.K.; Bhattacharjee, S.; Mahato, N.K. Process Parameters Optimization in Drilling Using Taguchi Method. In Proceedings of the Recent Trends in Product Design and Intelligent Manufacturing Systems, Singapore, 30–31 December 2021; pp. 701–708.
18. Aamir, M.; Giasin, K.; Tolouei-Rad, M.; Ud Din, I.; Hanif, M.I.; Kuklu, U.; Pimenov, D.Y.; Ikhlaq, M. Effect of Cutting Parameters and Tool Geometry on the Performance Analysis of One-Shot Drilling Process of AA2024-T3. *Metals* **2021**, *11*, 854. [CrossRef]
19. Kang, J.; Yao, E. Study On Burrs And Hole Quality Of Drilling AA2024 Plates Based On FEM And Experimental Investigation. *J. Appl. Sci. Eng.* **2022**, *26*, 911–921. [CrossRef]
20. Giasin, K.; Hodzic, A.; Phadnis, V.; Ayvar-Soberanis, S. Assessment of cutting forces and hole quality in drilling Al2024 aluminium alloy: Experimental and finite element study. *Int. J. Adv. Manuf. Technol.* **2016**, *87*, 2041–2061. [CrossRef]
21. Ashrafi, S.A.; Davoudinejad, A.; Barzani, M.M.; Salehi, P. Investigation into effect of tool wear on drilling force and surface finish while dry drilling aluminum 2024. *Adv. Mater. Res.* **2012**, *548*, 387–392. [CrossRef]
22. Kurt, M.; Bagci, E.; Kaynak, Y. Application of Taguchi methods in the optimization of cutting parameters for surface finish and hole diameter accuracy in dry drilling processes. *Int. J. Adv. Manuf. Technol.* **2009**, *40*, 458–469. [CrossRef]
23. Benchmarkabrasives. HSS VS Cobalt VS Carbide Drill Bits: What to Choose. Available online: https://benchmarkabrasives.com/blogs/news/hss-vs-cobalt-vs-carbide-drill-bits-what-to-choose (accessed on 30 June 2023).
24. Aamir, M.; Tolouei-Rad, M.; Giasin, K.; Vafadar, A. Feasibility of tool configuration and the effect of tool material, and tool geometry in multi-hole simultaneous drilling of Al2024. *Int. J. Adv. Manuf. Technol.* **2020**, *111*, 861–879. [CrossRef]
25. Oberg, E. *Machinery's Handbook 29th Edition-Full Book*; Industrial Press: New York, NY, USA, 2012.
26. Aamir, M.; Giasin, K.; Tolouei-Rad, M.; Vafadar, A. A review: Drilling performance and hole quality of aluminium alloys for aerospace applications. *J. Mater. Res. Technol.* **2020**, *9*, 12484–12500. [CrossRef]
27. Mohammadi, F.; Mirhashemi, M.; Rashidzadeh, R. A Coordinate Measuring Machine with Error Compensation in Feature Measurement: Model Development and Experimental Verification. *Int. J. Adv. Manuf. Technol.* **2022**. [CrossRef]
28. Mydin, N.M.M.; Dahnel, A.N.; Raof, N.A.; Khairussaleh, N.M.; Mokhtar, S. Comparison of Tool Wear Mechanisms During Drilling of Aluminium Alloy 7075 in Dry and Chilled Air Conditions. *J. Sci. Technol.* **2022**, *14*, 67–74. [CrossRef]

29. Musavi, S.H.; Sepehrikia, M.; Davoodi, B.; Niknam, S.A. Performance analysis of developed micro-textured cutting tool in machining aluminum alloy 7075-T6: Assessment of tool wear and surface roughness. *Int. J. Adv. Manuf. Technol.* **2022**, *119*, 3343–3362. [CrossRef]
30. Ramulu, M.; Branson, T.; Kim, D. A study on the drilling of composite and titanium stacks. *Compos. Struct.* **2001**, *54*, 67–77. [CrossRef]

machines

MDPI

Article

Optimizing Layer Thickness and Width for Fused Filament Fabrication of Polyvinyl Alcohol in Three-Dimensional Printing and Support Structures

Mahmoud Moradi [1,*], Mojtaba Karamimoghadam [2,*], Saleh Meiabadi [3], Shafqat Rasool [1], Giuseppe Casalino [2], Mahmoud Shamsborhan [4], Pranav Kattungal Sebastian [1], Arun Poulose [1], Abijith Shaiju [1] and Mohammad Rezayat [5]

1 Faculty of Arts, Science and Technology, University of Northampton, Northampton NN1 5PH, UK; shafqat.rasool@northampton.ac.uk (S.R.); pranavkattungal@gmail.com (P.K.S.); arunkoonathan@gmail.com (A.P.); abijithshaiju042@gmail.com (A.S.)

2 Department of Mechanics, Mathematics and Management, Polytechnic University of Bari, Via Orabona 4, 70125 Bari, Italy; giuseppe.casalino@poliba.it

3 Department of Mechanical Engineering, École de Technologie Supérieure, 1100 Notre-Dame West, Montreal, QC H3C 1K3, Canada; mohammadsaleh.sheikhmohammadmeiabadi.1@ens.etsmtl.ca

4 Department of Mechanical Engineering, University of Zakho, Kurdistan Region 42001, Iraq; m.shamsborhan@gmail.com

5 Center for Structural Integrity, Micromechanics, and Reliability of Materials (CIEFMA)-Department of Materials Science and Engineering, Universitat Politècnica de Catalunya-BarcelonaTECH, 08019 Barcelona, Spain; mohammad.rezayat@upc.edu

* Correspondence: mahmoud.moradi@northampton.ac.uk (M.M.); m.karamimoghadam@phd.poliba.it (M.K.)

Abstract: Polyvinyl Alcohol (PVA) is frequently applied as a support material in 3D printing, especially in the crafting of intricate designs and projecting elements. It functions as a water-soluble filament, often paired with materials like ABS or PLA. PVA serves as a momentary scaffold, supporting the jutting segments of a 3D model throughout the printing process. Subsequent to printing, the primary component can be effortlessly isolated by dissolving the PVA support using water. PVA, being a pliable and eco-friendly polymer, is susceptible to moisture. Its aqueous solubility renders it a prime selection for bolstering 3D print structures. In this investigation, equivalent-sized samples were 3D printed utilizing an Ultimaker 3D printer to assess the potency of PVA-generated specimens. Tensile examinations were executed on each sample employing a testing apparatus. The durability of the specimens was notably impacted by the input parameters, specifically the stratum width and stratum thickness. Strength dwindled as stratum width increased, whereas it rose with augmented stratum thickness. A few specimens with heightened stratum width and compromised quality displayed subpar performance during the tensile assessment. The findings unveiled a peak tensile strength of 17.515 MPa and a maximum load of 1600 N. Attaining an optimal degree of material utilization led to a decrease in filament consumption by 8.87 g, all the while upholding a MTS (maximum tensile strength) of 10.078 MPa.

Keywords: Fused Filament Fabrication (FFF); Polyvinyl Alcohol (PVA); layer thickness; width thickness; 3D printing; tensile test

1. Introduction

Three-dimensional printing, also known as additive manufacturing (AM), involves the construction of three-dimensional objects from digital models by depositing, connecting, or solidifying material under computer control [1–3]. Unlike traditional manufacturing methods that involve material removal, 3D printing builds objects by layering materials [4–7]. The printing process involved dual extrusion and fused filament fabrication techniques. The quality and precision of the printed object were found to be influenced by factors such as temperature management, material viscosity, layer adhesion, and the mechanical, thermal, and optical properties of the material [8–10]. Three-dimensional printing has revolutionized manufacturing by enabling rapid prototyping, shorter lead times, and the production of customized products. It has found applications in various industries, including healthcare, aerospace, defense, automotive, and education, facilitating innovation, efficiency, and experiential learning opportunities [11–14].

Sikidar et al. [15] investigated the effects of layer thickness on the mechanical characteristics of 3D-printed ABS polymer samples. They utilized a fused deposition modelling (FDM) 3D printing device to produce samples with varying layer thicknesses and compared them to a sample made through conventional injection molding. The results indicated that the samples produced via injection molding exhibited the highest values of tensile strength, impact strength, and hardness. Chaudhery et al. [16] comprehensively explored the 3D printing process, including methodology, materials, feed, technology, and applications. They introduced 3D/4D printing technologies and highlighted the use of PVA as a feedstock. The authors emphasized that traditional methods lacked the ability to produce complex structures, whereas 3D printing technologies, such as light-, droplet-, and extrusion-based systems, enabled their creation. The study also investigated the temperature response of PVA. Terranova et al. [17] investigated the synthesis of PVA-based nanocomposites using diamond nanograins as fillers and their application in 3D printing through additive manufacturing. They explored the use of PVA and detonation nanodiamond dispersions as innovative inks for layer-by-layer additive manufacturing of variously shaped objects. The researchers developed 3D printing technology and methods to shape hybrid materials while fabricating nanocomposites, utilizing PVA-DND inks as a test system. They emphasized the importance of aligning the chemical and physical properties of the materials with the 3D printer to enhance the quality of the final printed products. The study demonstrated that well-defined and shaped structures of PVA-DND nanocomposites can be successfully printed, offering potential applications across various fields. Dwiyati et al. [18] conducted research on the axial and lateral tensile characteristics of Acrylonitrile Butadiene Styrene (ABS) material for 3D printing. Specimens were printed with varying layer thicknesses of 0.1, 0.2, and 0.3 mm, following ASTM D 632-02 standards [19]. Tensile tests were performed using the Zwick Roell Series Z 021 machine, and SEM testing was used to analyze the fracture surface. The study concluded that the axial direction of 3D-printed specimens exhibited higher maximum force and tensile strength compared to the lateral direction, with thicker layers more likely to exhibit greater maximum force and tensile strength. Yu et al. [20] investigated the mechanical behavior of 3D preforms and their composites produced via additive manufacturing, focusing on the influence of printing direction. The compressive behavior of 3D braid preforms, and composites was analyzed for three different printing orientations (0°, 45°, and Z-direction). Pores induced by the fabrication process were observed in sections printed at 0° and 45° orientations. Solid cube specimens were then created and injected with a silicone matrix. Performs printed in the 45° direction exhibited improved inter-yarn adhesion, leading to enhanced initial modulus. However, Z-direction specimens displayed greater structural ductility due to inter-yarn slippage. Paul et al. [21] explored the application of novel biomaterials and advanced 3D printing techniques in bioprinting. The study highlights recent advancements in 3D printing technology and new materials, emphasizing their superiority over traditional methods, particularly in the field of biomedicine. Key considerations discussed include printing speed, feasibility of cell growth, and the ability to achieve complex shapes in bioprinting applications. Several studies have been conducted to investigate the mechanical properties of 3D-printed samples made from polyvinyl alcohol (PVA) and establish a correlation between printer parameters and mechanical characteristics [22–27]. These research efforts aim to understand how variations in printer settings, such as layer thickness, infill density, and printing speed, affect the resulting mechanical properties of PVA-based 3D-printed objects [28–32]. By identifying these relationships, researchers can optimize printer parameters to achieve desired mechanical properties in PVA 3D-printed samples [33–36]. AM, particularly Material Extrusion (MEX), refers to a swift and convenient manufacturing technique that employs raw materials in filament structure. It holds the potential to effectively utilize recycled plastics and fibers sourced from industrial waste and household recycling, making it a viable approach for sustainability [37,38].

This study explores various methods employed in 3D printing technologies, specifically utilizing a dual extrusion print head and fused filament fabrication. The objective is to investigate the influence of input parameters, including layer thickness and wall thickness, on the strength of specimens. To conduct the experiment, shape samples were produced using PVA as the material on an Ultimaker 3D printer, while adhering to the specified input parameters. Subsequently, measurements were taken from the output samples and subjected to a tensile test. Results were obtained through graphical analysis, providing insights into material strength and force analysis. Furthermore, optimization and design of experiments were applied, utilizing the response surface method to optimize the weight and maximum tensile strength of the 3D-printed samples. The goal is to enhance the weight efficiency and mechanical characteristics of PVA samples subsequent to the 3D printing procedure.

2. Experimental Work

The experiment aimed to use an Ultimaker 5 3D printer to construct samples and investigate the impact of layer thickness and width thickness on their mechanical properties. The process involved creating a 3D design using SolidWorks software (version 11) and converting it into a series of cross-sectional layers. Fused filament fabrication (FFF) technology was employed, where a molten material was layered to build the items. Thermoplastic polymers like ABS and Nylon were used, and a dual extrusion print head with an auto nozzle lifting system was utilized. PVA material was prepared by being feed into the printer's filament spool and then into the print head, and the parameters were defined based on the response surface method levels (Table 1). In this study, the selected infill pattern was the honeycomb design. The specimens were fabricated following the dimensions specified for the ISO 527-2 tensile test samples [39].

Table 1. Independent process parameters with design levels.

Variable	Notation	Unit	−2	−1	0	1	2
Layer thickness	LT	[mm]	0.1	0.15	0.2	0.25	0.3
Width thickness	WT	[mm]	0.4	0.6	0.8	1	1.2

Table 2 provides the chemical properties of PVA. The printing process involved extruding PVA layer by layer to form the shape of the object. In some cases, support structures were used and later removed. Once printing was completed, the product was removed from the print bed and any remaining support structures were taken out. The samples were then subjected to tensile testing to analyze their mechanical properties.

Table 2. Chemical properties of PVA.

Formula	(C2H40) x
Melting Point	200 °C
Density	1.19 g/cm^3
Boiling Point	228 °C
Solubility in	Water
EPA	DTXSID431930
Log P	0.26

After the completion of 3D printing, additional post-processing steps can be performed to improve the surface quality of the object, including sanding or polishing. The removal of PVA support material is achieved by submerging the printed object in water, causing the PVA to dissolve and leaving behind the primary 3D-printed component. Once dried, the part may require further post-processing, such as sanding or finishing, to achieve the desired outcome. The result is a 3D-printed object constructed by PVA materials. This procedure was repeated to produce a total of eleven samples, and the final products can be observed in Figure 1. Geometrical parameters, including weight, length, and thickness, were measured for each sample (Table 3). Subsequently, all the 3D-printed samples underwent tensile strength testing using an Instron 5567 Universal Testing Machine (UTM). The nozzle temperature was set within the range of 180 to 280 degrees Celsius with a nozzle diameter of 0.6 mm.

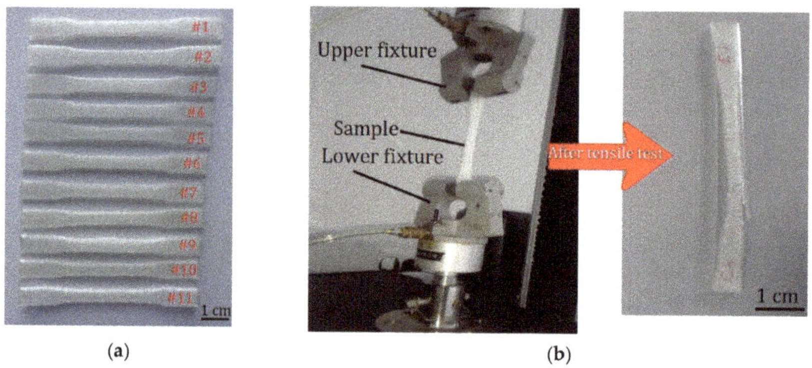

Figure 1. (**a**) Final products (3D-printed object made with PVA material). (**b**) The samples during and after tensile testing.

Table 3. Experimental layout and multi-performance results.

Experiment No.	Input Variables		Responses	
	LT [mm]	WT [mm]	MTS (MPA)	Weight (g)
1	0.2	1.2	8.892	8.34
2	0.25	0.6	13.878	9.26
3	0.2	0.4	17.515	9.40
4	0.2	0.8	12.708	9.58
5	0.2	0.8	14.175	9.68
6	0.1	0.8	13.431	9.46
7	0.3	0.8	14.507	9.47
8	0.15	0.6	17.925	9.68
9	0.15	1	10.458	8.83
10	0.2	0.8	14.587	9.40
11	0.25	1	4.417	7.95

3. Result and Discussion

3.1. Maximum Tensile Strength (MTS)

The ANOVA Table 4 presents the designed model for the MTS of the samples after 3D printing. In this model, the effects of input parameters on the MTS are analyzed. Based on the *p*-values and the significant model experiments, the effects of the parameters are deemed significant. Referring to the ANOVA Table 4 and the coefficients Equation (1), it can be concluded that the effects of the WT parameter on the MTS are greater compared to the effects of the LT parameter in the printed samples.

$$(MTS)^{2.26} = +1090.15231 - 560.87673 \times LT - 768.57399 \times WT \quad (1)$$

Table 4. Analysis of variance (ANOVA) for MTS.

Source	Sum of Squares	df	Mean Square	F Value	*p*-Value Prob > F
Model	2.930×10^5	2	1.465×10^5	13.39	0.0028
A-LT	9437.48	1	9437.48	0.86	0.3802
B-WT	2.835×10^5	1	2.835×10^5	25.91	0.0009
Residual	87530.92	8	10941.36		
Lack of Fit	80375.07	6	13395.84	3.74	0.2258
Pure Error	7155.85	2	3577.93		
Cor Total	3.805×10^5	10			

In Figure 2, the output graphs generated by the "Design Expert" software for the printed sample's MTS are displayed. Figure 2a represents the normal plot of residuals for the MTS, showing that the residuals align closely with the trend line, indicating the conformity of the DOE for this output. Figure 2b, which is the perturbation plot for the MTS, examines the simultaneous effects of two parameters, WT and LT, on the MTS of the samples. Figure 2c presents the surface plot for the MTS of the samples. Based on this plot and the 2D contour plot of the MTS (Figure 2d), it can be observed that the MTS increases with a decrease in the WT of the printer. This is because reducing the WT results in thinner printed layers, leading to better filament consolidation in the samples. Increased consolidation of the printed layers requires more force to separate the samples, ultimately resulting in an increase in the MTS. Furthermore, considering the red regions in Figure 2d, the MTS of the samples is maximized when the WT parameter is set to a lower value for the printer. This is because decreasing the WT increases the print density and filament consumption, which leads to an increase in the MTS of the samples.

The tensile test results were obtained using a computer-controlled tensile testing machine. The results were analyzed and presented in the form of graphs. In Figure 3a, the load applied to sample 1 is plotted against the corresponding extension. It is evident that the sample exhibits elastic behavior up to an extension of 1.75 mm with a force of 510 N. Beyond this point, the material deviates from its elastic state. The maximum load observed is 760 N, achieved at a stretch of 3 mm. Subsequently, as the sample is stretched further, a decrease in force is observed until fracture occurs at a 7 mm extension. Similarly, in Figure 3b, the load applied to sample 2 is plotted against the extension. The graph shows that the sample retains its elastic properties up to an extension of approximately 1.00 mm, with a force of around 500 N. Upon exceeding this limit, the material loses its elastic behavior. The maximum load recorded is 1050 N at an extension of 2.5 mm. As the sample is stretched beyond this point, a decrease in force is observed until the sample fractures at an extension of over 7.25 mm. Comparing the results of sample 1 and sample 2, it is noteworthy that sample 2, which was printed with a layer thickness of 0.25 mm and width thickness of 0.6 mm, exhibited a 36 percent increase in maximum load and a 56 percent increase in the MTS of Elasticity compared to sample 1. Furthermore, the width thickness of sample 2 was reduced to half of that in sample 1.

3.2. Weight

In ANOVA, Table 5, the designed model for the weight of the samples after 3D printing is presented. In this model, the effects of the input parameters on the weight of the samples, as well as the second-order effects of WT, are analyzed. Based on the p-values, the effects of the parameters and the experimental model are found to be significant. Considering ANOVA, Table 5, and the coefficients Equation (2), it can be concluded that the effects of the WT parameter on the weight of the printed samples are greater compared to the effects of the LT parameter.

$$(\text{Weight})^3 = +655.92146 - 489.41132 \times \text{LT} + 1087.08498 \times \text{WT} - 942.73992 \times \text{WT}^2 \quad (2)$$

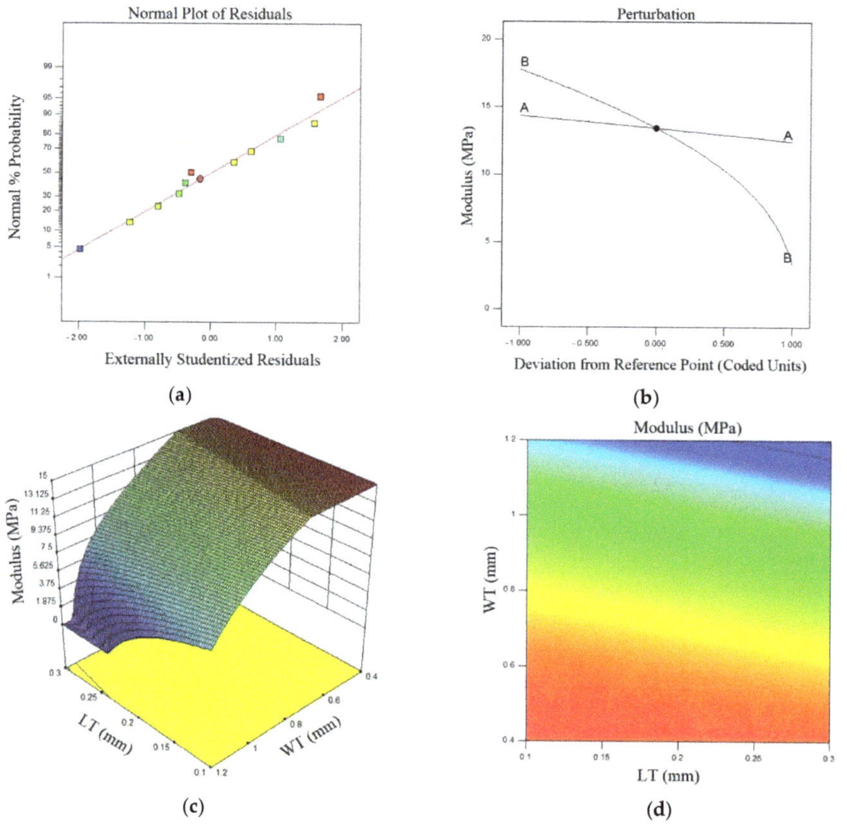

Figure 2. (**a**) Normal plot of residuals of MTS (**b**) Perturbation plot of MTS (**c**) 3D surface plot of MTS in terms of LT and WT (**d**) contour plot of MTS.

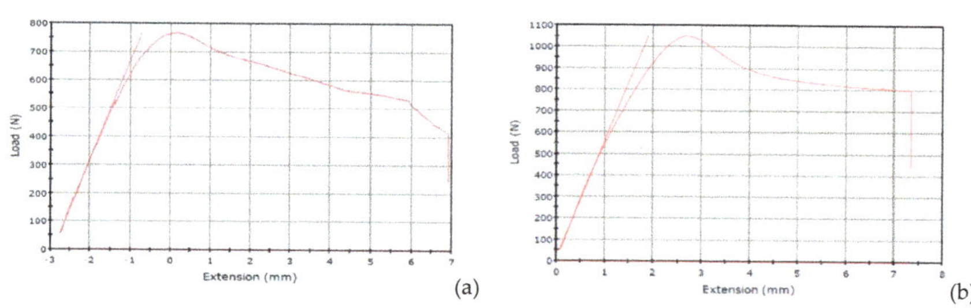

Figure 3. (**a**) Load (N) vs. Extension (mm) of Sample 1 (**b**) Load (N) vs Extension (mm) of Sample 2.

In Figure 4, the output graphs of the Design Expert software (version 11)for the weight of the printed samples are shown. In Figure 4a, the normal plot of residuals for weight is displayed, indicating the residuals align closely to the diagonal line, indicating the adequacy of the DOE for this output. In Figure 4b, the surface plot for weight is presented. Based on this plot and the 2D contour plot of weight (Figure 4c), it can be observed that increasing the weight of the samples is directly related to an increase in LT. This is because an increase in LT leads to thicker printed layers, resulting in higher filament consumption and ultimately increasing the weight of the samples. Additionally, considering the red regions in Figure 4c, the weight of the samples is minimized when the WT parameter is low and the LT parameter is set on high for the printer, as reducing WT increases print density and filament consumption, leading to an increase in sample weight.

Table 5. Analysis of variance (ANOVA) for Weight.

Source	Sum of Squares	Df	Mean Square	F Value	*p*-Value Prob > F
Model	1.250×10^5	3	41,653.09	5.12	0.0348
A-LT	7185.70	1	7185.70	0.88	0.3787
B-WT	85,196.52	1	85,196.52	10.47	0.0143
B^2	32,577.04	1	32,577.04	4.00	0.0855
Residual	56,966.61	7	8138.09		
Lack of Fit	53,971.72	5	10,794.34	7.21	0.1263
Pure Error	2994.90	2	1497.45		
Cor Total	1.819×10^5	10			

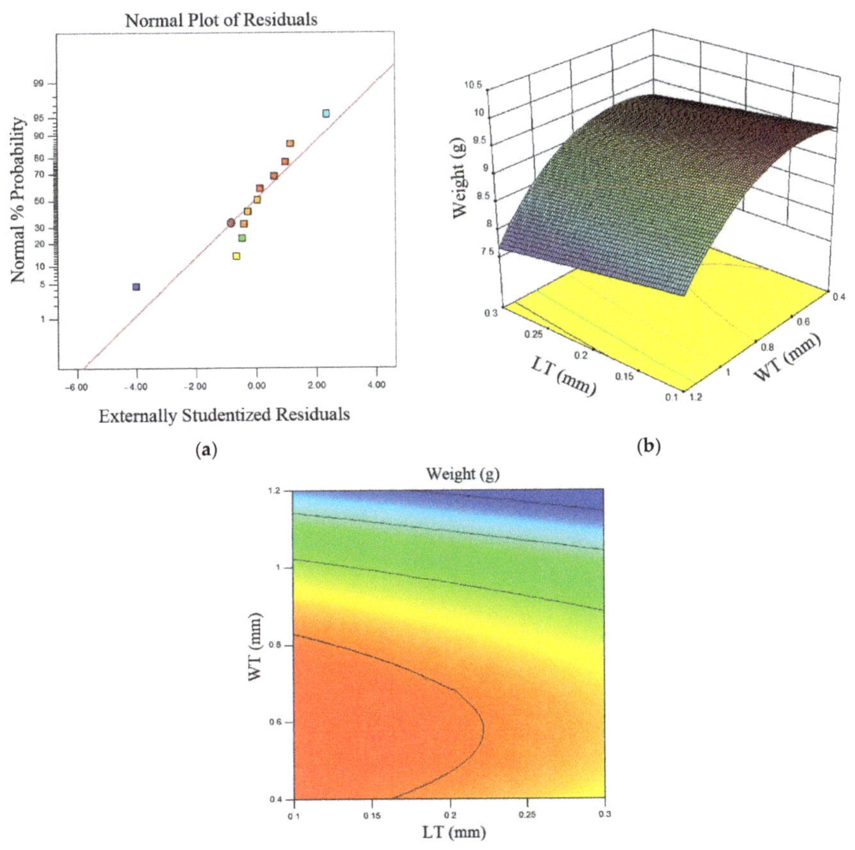

Figure 4. (**a**) Normal plot of residuals of Weight; (**b**) 3D surface plot of Weight in terms of LT and WT (**c**) contour plot of Weight.

4. Optimization

In this section, the optimization of 3D-printed PVA samples using the FFF method is addressed. In this optimization, two input parameters of the printer, namely WT and LT, were optimized based on the optimization Table 6, where the parameter ranges for both samples were evaluated, considering the upper and lower limits. Since the responses for this study are MTS and Weight, a goal was set for each in the optimization analysis, aiming to increase the MTS and decrease the Weight of the printed PVA samples. Furthermore, according to Table 7, three optimized samples were shown for this experiment, and the predicted values by the Design Expert software and the actual data from the tests were reported for these three optimized samples. As evident, the error values for the samples are below 15%, which is highly suitable for predicting the optimization of the samples based on previous studies, and conducted research. Additionally, considering the overlay plot in Figure 5, it demonstrates the optimized regions for the input parameters of the printer, where selecting a parameter within the yellow regions will result in optimal responses.

Table 6. Constraints and criteria of input parameters and responses.

Name	Goal	Lower Limit	Upper Limit	Lower Weight	Upper Weight	Importance
A: LT	is in range	0.1	0.3	1	1	3
B: WT	is in range	0.4	1.2	1	1	3
MTS	maximize	4.41779	17.925	1	1	3
Weight	minimize	7.95	9.68	1	1	3

Table 7. Predicted optimum results and experimental validation.

No.	LT	WT	Desirability		MTS (MPa)	Weight (g)
#1	0.3	0.85	1	Predicted	13.540	9.15
				Actual	11.985	9.05
				Error %	12.97	1.10
#2	0.25	0.85	0.953	Predicted	11.603	8.58
				Actual	12.091	8.99
				Error %	4.03	4.56
#3	0.25	0.95	0.947	Predicted	10.815	9.07
				Actual	10.078	8.87
				Error %	7.31	2.25

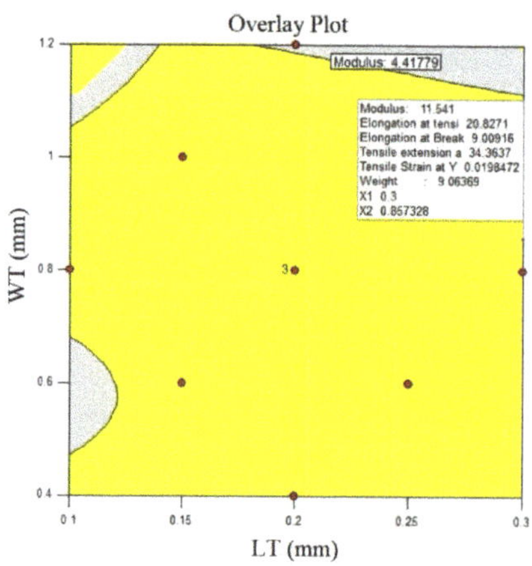

Figure 5. Overlay plots in terms of LT and WT.

5. Conclusions

The constraints and future potential of PVA 3D printing encompass its susceptibility to moisture, causing degradation, print issues, and extruder clogs, necessitating proper storage. Primarily used as support material in dual-extrusion 3D printing, PVAs limited application excludes standalone objects due to properties and cost. Slower print speeds increase project duration, deterring time-sensitive tasks. Higher cost hampers larger projects and cost-effectiveness goals. Its biocompatibility is promising for temporary medical implants. PVA finds utility in education and rapid prototyping due to easy support removal. Anticipated technological advancements could alleviate limitations and expand prospects as researchers strive to optimize PVA-based 3D printing for diverse applications. Below are notable accomplishments from this research:

- Mechanical properties of printed samples were significantly influenced by the width thickness parameter.
- Low layer thickness and high width thickness combo reduces tensile strength due to poor layer adhesion in width direction.
- Using 0.25 mm layer thickness and 1 mm width thickness requires less material but increases voids and lowers tensile strength.
- Sample 8: 0.15 mm layer thickness, 0.6 mm width thickness, MTS of 17.925 MPa.
- Sample 3: 0.2 mm layer thickness, 0.4 mm width thickness, MTS of 17.515 MPa, maximum load 1600 N. Sample 3 maintains good MTS with a reduced filament consumption of 8.87 g (10.078 MPa).

Author Contributions: Conceptualization, M.M. and M.K.; methodology, M.M., M.K., S.M., G.C., S.R., M.S. and M.R.; software, S.M.; validation, P.K.S., A.P., S.M. and M.S.; formal analysis, S.M.; investigation, M.M., A.S. and G.C.; Experimental work and data curation, P.K.S., S.R., A.P., A.S. and M.M.; writing—original draft preparation, M.K. and S.M.; writing—review and editing, M.M., M.K., G.C., M.S. and M.R.; visualization, M.M., M.K., S.M., G.C., M.S. and M.R.; supervision, M.M., G.C. and M.K. All authors have read and agreed to the published version of the manuscript.

Funding: This research received no external funding.

Data Availability Statement: Data detail is already provided in the contents of this paper.

Conflicts of Interest: The authors declare no conflict of interest.

Abbreviations

FFF	Fused filament fabrication
DOE	Design of Experiment
RSM	Response Surface Methodology
PVA	Polyvinyl Alcohol
ANOVA	Analysis of Variance
ABS	Acrylonitrile Butadiene Styrene
LT	Layer thickness
WT	Width thickness
UTM	Universal Testing Machine
MTS	Maximum Tensile Strength

References

1. Nazir, A.; Gokcekaya, O.; Billah, K.M.M.; Ertugrul, O.; Jiang, J.; Sun, J.; Hussain, S. Multi-material additive manufacturing: A systematic review of design, properties, applications, challenges, and 3D Printing of materials and cellular metamaterials. *Mater. Des.* **2023**, *226*, 111661. [CrossRef]
2. Moradi, M.; Karami Moghadam, M.; Shamsborhan, M.; Bodaghi, M. The synergic effects of FDM 3D printing parameters on mechanical behaviors of bronze poly lactic acid composites. *J. Compos. Sci.* **2020**, *4*, 17. [CrossRef]
3. Ouassil, S.E.; El Magri, A.; Vanaei, H.R.; Vaudreuil, S. Investigating the effect of printing conditions and annealing on the porosity and tensile behavior of 3D-printed polyetherimide material in Z-direction. *J. Appl. Polym. Sci.* **2023**, *140*, e53353. [CrossRef]

4. Al-Dulaijan, Y.A.; Alsulaimi, L.; Alotaibi, R.; Alboainain, A.; Akhtar, S.; Khan, S.Q.; Al-Ghamdi, M.; Gad, M.M. Effect of Printing Orientation and Postcuring Time on the Flexural Strength of 3D-Printed Resins. *J. Prosthodont.* **2023**, *32*, 45–52. [CrossRef] [PubMed]
5. Kara, Y.; Kovács, N.K.; Nagy-György, P.; Boros, R.; Molnár, K. A novel method and printhead for 3D printing combined nano-/microfiber solid structures. *Addit. Manuf.* **2023**, *61*, 103315. [CrossRef]
6. Meiabadi, M.S.; Moradi, M.; Karamimoghadam, M.; Ardabili, S.; Bodaghi, M.; Shokri, M.; Mosavi, A.H. Modeling the producibility of 3D printing in polylactic acid using artificial neural networks and fused filament fabrication. *Polymers* **2021**, *13*, 3219. [CrossRef]
7. Tomal, W.; Kiliclar, H.C.; Fiedor, P.; Ortyl, J.; Yagci, Y. Visible light induced high resolution and swift 3D printing system by halogen atom transfer. *Macromol. Rapid Commun.* **2023**, *44*, 2200661. [CrossRef]
8. An, Z.; Liu, Z.; Mo, H.; Hu, L.; Li, H.; Xu, D.; Chitrakar, B. Preparation of Pickering emulsion gel stabilized by tea residue protein/xanthan gum particles and its application in 3D printing. *J. Food Eng.* **2023**, *343*, 111378. [CrossRef]
9. Dey, D.; Panda, B. An experimental study of thermal performance of 3D printed concrete slabs. *Mater. Lett.* **2023**, *330*, 133273. [CrossRef]
10. Moradi, M.; Karami Moghadam, M.; Shamsborhan, M.; Bodaghi, M.; Falavandi, H. Post-processing of FDM 3D-printed polylactic acid parts by laser beam cutting. *Polymers* **2020**, *12*, 550. [CrossRef]
11. Zhang, X.N.; Zheng, Q.; Wu, Z.L. Recent advances in 3D printing of tough hydrogels: A review. *Compos. Part B Eng.* **2022**, *238*, 109895. [CrossRef]
12. Zhong, H.; Zhang, M. 3D printing geopolymers: A review. *Cem. Concr. Compos.* **2022**, *128*, 104455. [CrossRef]
13. Zhu, Y.; Tang, T.; Zhao, S.; Joralmon, D.; Poit, Z.; Ahire, B.; Keshav, S.; Raje, A.R.; Blair, J.; Zhang, Z.; et al. Recent advancements and applications in 3D printing of functional optics. *Addit. Manuf.* **2022**, *52*, 102682. [CrossRef]
14. Zhang, F.; Li, Z.; Xu, M.; Wang, S.; Li, N.; Yang, J. A review of 3D printed porous ceramics. *J. Eur. Ceram. Soc.* **2022**, *42*, 3351–3373. [CrossRef]
15. Shubham, P.; Sikidar, A.; Chand, T. The influence of layer thickness on mechanical properties of the 3D printed ABS polymer by fused deposition modeling. *Key Eng. Mater.* **2016**, *706*, 63–67. [CrossRef]
16. Mallakpour, S.; Tabesh, F.; Hussain, C.M. A new trend of using poly (vinyl alcohol) in 3D and 4D printing technologies: Process and applications. *Adv. Colloid Interface Sci.* **2022**, *301*, 102605. [CrossRef]
17. Angjellari, M.; Tamburri, E.; Montaina, L.; Natali, M.; Passeri, D.; Rossi, M.; Terranova, M.L. Beyond the concepts of nanocomposite and 3D printing: PVA and nanodiamonds for layer-by-layer additive manufacturing. *Mater. Des.* **2017**, *119*, 12–21. [CrossRef]
18. Dwiyati, S.T.; Kholil, A.; Riyadi, R.; Putra, S.E. Influence of layer thickness and 3D printing direction on tensile properties of ABS material. *J. Phys. Conf. Ser.* **2019**, *1402*, 066014. [CrossRef]
19. Rajpurohit, S.R.; Dave, H.K. Tensile Properties of 3D Printed PLA under Unidirectional and Bidirectional Raster Angle: A Comparative Study. *Int. J. Mater. Metall. Eng.* **2018**, *22*, 6–11.
20. Quan, Z.; Suhr, J.; Yu, J.; Qin, X.; Cotton, C.; Mirotznik, M.; Chou, T.W. Printing direction dependence of mechanical behavior of additively manufactured 3D preforms and composites. *Compos. Struct.* **2018**, *184*, 917–923. [CrossRef]
21. Pradeep, P.V.; Paul, L. Review on novel biomaterials and innovative 3D printing techniques in biomedical applications. *Mater. Today Proc.* **2022**, *58*, 96–103. [CrossRef]
22. Stammen, J.A.; Williams, S.; Ku, D.N.; Guldberg, R.E. Mechanical properties of a novel PVA hydrogel in shear and unconfined compression. *Biomaterials* **2001**, *22*, 799–806. [CrossRef] [PubMed]
23. Kamran, M.; Saxena, A. A comprehensive study on 3D printing technology. *MIT Int. J. Mech. Eng.* **2016**, *6*, 63–69.
24. Hema, M.; Selvasekerapandian, S.; Hirankumar, G.; Sakunthala, A.; Arunkumar, D.; Nithya, H. Structural and thermal studies of PVA: NH4I. *J. Phys. Chem. Solids* **2009**, *70*, 1098–1103. [CrossRef]
25. Markiz, N.; Horváth, E.; Ficzere, P. Influence of printing direction on 3D printed ABS specimens. *Prod. Eng. Arch.* **2020**, *26*, 127–130. [CrossRef]
26. Noushini, A.; Samali, B.; Vessalas, K. Effect of polyvinyl alcohol (PVA) fibre on dynamic and material properties of fibre reinforced concrete. *Constr. Build. Mater.* **2013**, *49*, 374–383. [CrossRef]
27. Prabhakar, M.M.; Saravanan, A.K.; Lenin, A.H.; Mayandi, K.; Ramalingam, P.S. A short review on 3D printing methods, process parameters and materials. *Mater. Today Proc.* **2021**, *45*, 6108–6114. [CrossRef]
28. Liu, B.; Liu, X.; Li, G.; Geng, S.; Li, Z.; Weng, Y.; Qian, K. Study on anisotropy of 3D printing PVA fiber reinforced concrete using destructive and non-destructive testing methods. *Case Stud. Constr. Mater.* **2022**, *17*, e01519. [CrossRef]
29. Teodorescu, M.; Bercea, M.; Morariu, S. Biomaterials of PVA and PVP in medical and pharmaceutical applications: Perspectives and challenges. *Biotechnol. Adv.* **2019**, *37*, 109–131. [CrossRef]
30. Struzziero, G.; Barbezat, M.; Skordos, A.A. Assessment of the benefits of 3D printing of advanced thermosetting composites using process simulation and numerical optimisation. *Addit. Manuf.* **2023**, *63*, 103417. [CrossRef]
31. Wang, Q.; Zhang, G.; Zheng, X.; Ni, Y.; Liu, F.; Liu, Y.; Xu, L.R. Efficient characterization on the interlayer shear strengths of 3D printing polymers. *J. Mater. Res. Technol.* **2023**, *22*, 2768–2780. [CrossRef]
32. Aki, D.; Ulag, S.; Unal, S.; Sengor, M.; Ekren, N.; Lin, C.C.; Yılmazer, H.; Ustundag, C.B.; Kalaskar, D.M.; Gunduz, O. 3D printing of PVA/hexagonal boron nitride/bacterial cellulose composite scaffolds for bone tissue engineering. *Mater. Des.* **2020**, *196*, 109094. [CrossRef]

33. Hua, W.; Shi, W.; Mitchell, K.; Raymond, L.; Coulter, R.; Zhao, D.; Jin, Y. 3D printing of biodegradable polymer vascular stents: A review. *Chin. J. Mech. Eng. Addit. Manuf. Front.* **2022**, *1*, 100020. [CrossRef]
34. Baniasadi, H.; Madani, Z.; Ajdary, R.; Rojas, O.J.; Seppälä, J. Ascorbic acid-loaded polyvinyl alcohol/cellulose nanofibril hydrogels as precursors for 3D printed materials. *Mater. Sci. Eng. C* **2021**, *130*, 112424. [CrossRef]
35. Rovetta, R.; Pallavicini, A.; Ginestra, P.S. Bioprinting process optimization: Case study on PVA (Polyvinyl Alcohol) and Graphene Oxide biocompatible hydrogels. *Procedia CIRP* **2022**, *110*, 145–149. [CrossRef]
36. Moradi, M.; Karamimoghadam, M.; Meiabadi, S.; Casalino, G.; Ghaleeh, M.; Baby, B.; Ganapathi, H.; Jose, J.; Abdulla, M.S.; Tallon, P.; et al. Mathematical modelling of fused deposition modeling (FDM) 3D printing of poly vinyl alcohol parts through statistical design of experiments approach. *Mathematics* **2023**, *11*, 3022. [CrossRef]
37. Sam-Daliri, O.; Ghabezi, P.; Flanagan, T.; Finnegan, W.; Mitchell, S.; Harrison, N.M. Recovery of Particle Reinforced Composite 3D Printing Filament from Recycled Industrial Polypropylene and Glass Fibre Waste. *Proc. World Congr. Mech. Chem. Mater. Eng.* **2022**, *177*, 3–4.
38. Ghabezi, P.; Flanagan, T.; Harrison, N. Short basalt fibre reinforced recycled polypropylene filaments for 3D printing. *Mater. Lett.* **2022**, *326*, 132942. [CrossRef]
39. Yao, T.; Deng, Z.; Zhang, K.; Li, S. A method to predict the ultimate tensile strength of 3D printing polylactic acid (PLA) materials with different printing orientations. *Compos. Part B Eng.* **2019**, *163*, 393–402. [CrossRef]

machines

MDPI

Article

Manufacturability-Based Design Optimization for Directed Energy Deposition Processes

Harry Bikas, Michail Aggelos Terzakis and Panagiotis Stavropoulos *

Laboratory for Manufacturing Systems & Automation, University of Patras, 26504 Patras, Greece; bikas@lms.mech.upatras.gr (H.B.); terzakis@lms.mech.upatras.gr (M.A.T.)
* Correspondence: pstavr@lms.mech.upatras.gr; Tel.: +30-2610-910160

Abstract: Additive Manufacturing (AM) is the process of joining materials by selectively depositing them layer upon layer for the purpose of manufacturing parts or assemblies which are based on a 3D digital model. The nature of these processes results in the morphing of complex component geometries, enabling a high degree of design freedom and resulting in lightweight structures with increased performance. These processes, however, experience many limitations regarding manufacturability. The aim of this study is to develop a method and tool that optimizes the design of a component to avoid overhanging geometries and the need for supports during the Additive Manufacturing process. A workflow consisting of steps for topology optimization, orientation optimization, material addition, and machine code generation is described and implemented using Rhinoceros 3D and Grasshopper software. The proposed workflow is compared to a conventional workflow regarding manufacturing Key Performance Indicators (KPIs) such as part volume, support volume, and build time. A significant reduction is observed regarding all the KPIs by using the proposed method. Examining the results from both the conventional workflow and the proposed one, it is clear that the latter has unquestionable advantages in terms of effectiveness. In the particular case study presented, a total volume reduction of around 80% is observed. The reduction in the total volume (including the required support volume) leads to a significant reduction in the material used as well as in the build time, consequently resulting in cost reduction.

Keywords: additive manufacturing; DED; manufacturability; topology optimization; orientation optimization; material addition; slicing; G-code

1. Introduction

The process of building up a part by depositing material layer by layer on a build bed or within a build chamber is known as Additive Manufacturing (AM) [1]. Due to the freedom to create complicated geometries without the use of specialized tools, AM processes introduce an entirely novel realm within engineering design and manufacturing. These advantages have made Additive Manufacturing appealing to numerous industries, including automotive, aerospace, machine tool manufacturing, healthcare, and the food industry [2]. One of the differentiating elements between the various AM process families is the process mechanism used. This is a key aspect as the process mechanism can be linked directly with the manufacturability limitations of the process. As such, categorizing AM processes according to the process mechanism is useful [3]. From the plethora of AM processes available, this work focuses specifically on the Directed Energy Deposition (DED) process family. In DED, materials are fed through a narrow nozzle and are melted during deposition using focused thermal energy. The energy source can be a laser beam, an electron beam, or a plasma arc. The deposited materials are typically metals in the form of wire or powder [4]. Powder-based DED processes typically use a laser as the thermal energy source, while wire-based DED processes utilize an electric arc, a plasma arc, a laser beam, or an electron beam as the thermal energy source [5].

This paper presents a method and software tool that enhances the manufacturability of a component for DED processes that has been generated from a topology optimization procedure by limiting and modifying the geometry of overhanging areas and thus the required support volume to be built. The goal of the presented method and tool is to improve the manufacturability of highly optimized lightweight structural parts typically designed using topology optimization, reducing the total volume of material that needs to be deposited to manufacture both the part and the support structures, thus reducing material usage, build time, and, ultimately, costs.

The emergence of Additive Manufacturing technologies and their competitive advantage, in regard to the geometrical freedom that can be obtained, have led to the increased popularity of topology optimization, a method that makes lighter and potentially equally strong components by reducing unneeded material from the initial design [6]. Different algorithmic approaches have been proposed to implement topology optimization. Bendsoe et al. [7,8] presented a homogenization method that exploits infinitely microscale holes in a design domain rather than deleting the whole of a finite element [9–13]. Solid Isotropic Material with Penalization (SIMP) is another topology optimization method that penalizes the stiffness tensor by a penalization factor (p) that helps force a solution of either a solid or a void. A penalization factor greater than one turns the intermediate densities into solid and void zones, while a penalization value equal to one will result in a density gradient [14–18]. Besides density-based approaches, researchers inspired by the nature, specifically by the optimal, organic structures, of shells, bones, and trees found that their long evolutionary period and adaption to any environment they find themselves in are the reasons for such geometries. This led them to develop a new topology optimization method, called Evolutionary Structural Optimization (ESO). The ESO approach uses a fixed model with typical finite elements to describe the initial design domain. The ultimate optimal structure is a subset of the initial set of finite elements [19–22]. Based on the ESO technique, Yang et al. [23] developed an extension dubbed bidirectional ESO (BESO) for topology optimization sensitive to stiffness and displacement limitations.

All aforementioned topology optimization strategies produce objects with unique, organic shapes aiming to be manufactured utilizing the increasingly available AM processes. Nevertheless, even the most advanced AM processes have finite manufacturing capabilities; therefore, an evaluation of the manufacturing process limitations against the part to be additively manufactured should be conducted in order to attain a high level of manufacturability. Manufacturability in general encompasses a parts' design ease of manufacturing and its potential for cost reduction [2]. Similarly, AM manufacturability is the degree to which a part that is to be manufactured using an AM process will employ the favorable properties of this given technique [24]. In order to achieve a high degree of AM manufacturability, specific Design for AM (DfAM) rules must be taken into consideration during the early design stages of a part. Bikas et al. [25] underlined the importance of these guidelines, since following them will allow the part to fulfill the standards it is designed to and attain the optimum degree of manufacturability. Furthermore, they classified them into two groups, which are the design aspects (described as any characteristic which can be measured at the design stage, such as overhangs, bores, channels, etc., as well as a part's programming parameters) and the design considerations (the consequence on the manufactured part).

One of the most essential design aspects that significantly affects the manufacturing process are the overhangs since they are directly correlated with the need to support structures [26,27]. An overhanging feature is any shape whose orientation is not parallel to the build vector, contributing to the reduction in the efficiency of the AM process build time, material, post-processing equipment, and process cost. In their work, Ghiasian et al. [28] presented a framework to evaluate manufacturing feasibility using AM considering limits imposed from parameters such as geometry, build orientation/generated supports, required time, and cost. Thereupon, numerous methods based on computational methodologies that are focusing on tackling manufacturability challenges have been offered [29–31]. Lianos et al. [32] provide a paradigm that analyzes the manufacturability of a given design considering a Directed Energy Deposition (DED) technique. Subsequently, an AM design workflow is proposed, starting with the determination of the stress situation inside the build volume of the part and then orienting in order to minimize the overhanging features and maximize its AM manufacturability.

A small amount of works have also been proposed to manipulate existing design geometries with the aim of improving manufacturability. O'Hara et al. [33] proposed three advanced mesh manipulation techniques, and post-print surface optimization can decrease design time and improve physiological simulations, aiming to help plan surgical treatments and contribute to the success of test devices.

All these approaches that address manufacturability limitations are problems that need to optimize two or more parameters simultaneously under various constraints. These problems are called constrained multi-objective optimization problems (CMOPs), and their solution depends on evolutionary algorithms [34,35].

The last piece of this process is the development of the component that occurred from the whole optimization procedure. This includes the slicing of the fully optimized mesh and the generation of the appropriate machine code (G-code or otherwise) [36,37].

After studying the aforementioned works, the conclusion is that there is no software tool available that is able to optimize the geometry of a component aiming to eliminate the areas that reduce AM manufacturability. This work focuses on addressing this gap by utilizing the significant advantages of algorithmic-aided design to propose a design strategy based on manufacturability.

2. Materials and Methods

The proposed method consists of 4 steps that are linked together to generate an optimal (from an AM manufacturability perspective) design. Step 1 focuses on the topology optimization process for a given boundary domain where the structural analysis is going to be performed. Step 2 presents the orientation optimization process where the optimized mesh that occurs from Step 1 is rotated around the x- and y-axis in order to minimize the overhanging areas of the mesh and thus the required support material volume. Step 3 is the material addition process, which utilizes the mesh from the previous step to create suitable geometries with the aim of removing the remaining overhanging areas that are outside the feasible limits of the used AM process. The final step slices the fully optimized mesh from Step 3 to generate the machine code to manufacture the part. The aforementioned workflow is summarized in Figure 1.

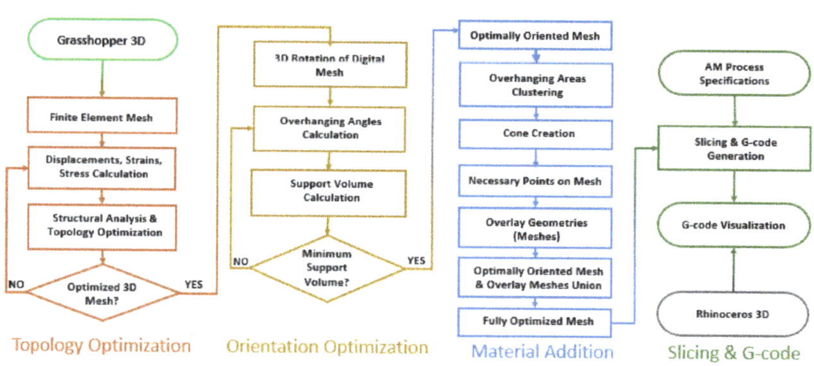

Figure 1. Method flow diagram.

The proposed workflow has been implemented utilizing the functionalities of the Rhinoceros 3D and Grasshopper software tools. Rhinoceros 3D is a computational-aided design-based software that uses NURBS-based tools in order to mathematically model and represent 2D or 3D geometries [38,39]. Grasshopper 3D is a visual programming language that uses a block diagram environment so as to perform numerous functions. It is a plug-in for Rhinoceros 3D, and it has the capability to run within it and can deploy its interface in order to visualize the 3D voxelized models [40,41]. The following sections present in detail each of the aforementioned algorithm steps.

2.1. Topology Optimization

Step 1 of the proposed method begins by defining the domain (boundary domain) in which the 3D volumetric analysis and optimization will be executed. The domain of the problem describes the region where the topology optimization operation will occur and the final component, consisting of a 3D ISO mesh, will be produced. Following this, the areas where the forces are applied and the constraints are installed must be specified in order to perform the intended operations. Then, the maximum number of iterations that the optimization algorithm will perform (the larger the number of iterations, the better the voxelized mesh), the percentage of material that will remain inside the boundary domain at the end of the procedure (it takes values between 0 and 1), and the penalty factor that prevents the greyscale result at the voxelized component (values 2 or 3 return the best problem covering) are determined to begin the topology optimization process.

2.2. Orientation Optimization

Step 2 of the proposed method deals with optimizing the orientation of the designed component to minimize overhangs. The starting point of this procedure is the determination of the overhanging areas of the optimized mesh. To do that, the angle between the normal vector of each triangular mesh facet and the global horizontal plane (XY plane) is calculated [42]. If this angle surpasses the limit that the respective AM process can create without supports, then this face is characterized as overhanging geometry [43]. The required support volume for the manufacturing process is computed for every triangular facet via the projection of the area of the overhanging geometry along the XY plane multiplied by their distance from the base plane. Aiming to minimize the required support volume, the optimal combination of rotation angles around the X- and Y-axis has to be determined. Searching every possible angle combination is time-consuming (and, depending also on the step size, possibly impractical). In order to tackle this multi-parameter optimization problem, an evolution algorithm is implemented. The algorithm follows an iteration process by dividing the range of angles into smaller ranges with a higher resolution when finding a better solution each time until it reaches the optimal one. The evolutionary algorithm is implemented through a Grasshopper component called "Galapagos" [44].

2.3. Material Addition

The next step in the proposed method aims to cover all the remaining overhanging areas with additional material to reduce or even eliminate the need for support structures in the manufacturing process. The general idea behind the material addition procedure is based on creating a geometry that complies with the manufacturing limitations of the AM process. The basic geometrical shape that its entire surface maintains a constant angle throughout its height is a cone. Therefore, the desired overlay geometry occurs from the Boolean difference in a conical surface that has a θ-degree normal vector (where θ is the limit imposed by the AM process) with a geometry that has been created from a set of points that define the border around the problematic overhanging area. After creating the conical surface, the next step of the procedure is to find the necessary points of the part to project on the aforementioned conical surface in order to create a new geometry that will cover the overhanging areas. At first, the points of the cluster are used to find their closest points on the conical surface. Then, these points (of the conical surface) are used to find a certain number of corresponding points of the initial mesh, and these, in turn, are used to locate more points on the conical surface (Figure 2a). It has to be noted that all the overlay geometries are separate meshes and are joined into a single mesh for the next step (Figure 2b).

(**a**) (**b**)

Figure 2. (**a**) Set of points and cone on mesh; (**b**) creation of overlay geometry on mesh.

2.4. Machine Code Generation

The last step of the proposed method is to slice the resulting geometry and generate the part program/machine code that is going to be used to build the part. This step is entirely dependent on the AM process that is going to be used to manufacture the optimized mesh that occurred from the previous process. In particular, parameters such as layer height, track width, deposition speed, etc., depend on the AM process and machine that will be used and determine the values that need to be entered towards generating the part program. The process parameters and variables that were used for the slicing and subsequent generation of the part program are summarized in Table 1.

Table 1. Process parameters.

Parameter	Value
Layer height [mm]	0.2
Infill [%]	30
Nozzle diameter [mm]	0.4
Shell number	2
Bottom layers	8
Top layers	0
Print speed [mm/s]	70
Travel speed [mm/s]	130
Retraction speed [mm/s]	30
Retractions distance [mm]	3
Filament diameter [mm]	1.75
Flow rate [%]	100

At first, the optimized mesh is sliced using an even layer height determined by the user based on the capabilities of the selected AM process. Then, the contours of the sliced mesh are utilized in order to create the infill pattern. The infill is measured as a percentage and expresses the area ratio of the layer enclosed by the contour that is going to be covered. This parameter is also determined by the user. Then, the contour and its respective infill curve are translated into points which are sorted in order to form a path that the process is going to follow to manufacture the part. The entirety of all these point coordinates, combined with the type of interpolation from point to point as well as the commands regarding the key functions of the process and machine, form the part program. This can be compiled as machine-specific code.

2.5. Case Study

In order to validate the proposed method, a case study of a simple cantilever beam was examined. The beam has initial dimensions of 300 × 200 × 120 (mm) for its length (x-axis), width (y-axis), and height (z-axis), respectively. A fixed constraint is applied in its left side, whereas a uniformly distributed linear load is applied throughout the y-axis, in an upward direction, in the lower right edge of the structure with a value of 200 N/m. (Figure 3a). The material is assumed to be 316L stainless steel with Young's Modulus E = 210 GPa and Poisson's Ratio ν = 0.3 [45]. It is worth mentioning that the properties used for the case study have been obtained from the literature and correspond to test coupons that were experimentally characterized using the same DED machine and material the authors intended to use [46]. These correspond to the properties directly obtained by the DED process without any post-processing, as this can often be the case for DED end users (contrary to the common practice of post processing in PBF processes). Since properties are dependent on the direction (and are typically lower across the deposition direction), the authors elected to use the worst scenario due to the orientation variation that characterizes the proposed method. The final material properties used are summarized in Table 2.

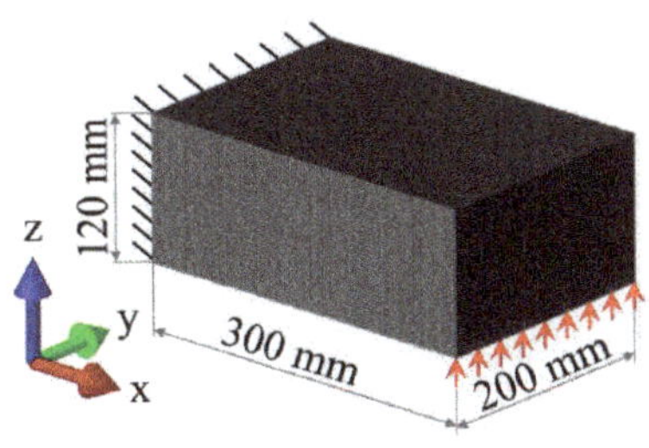

(a)

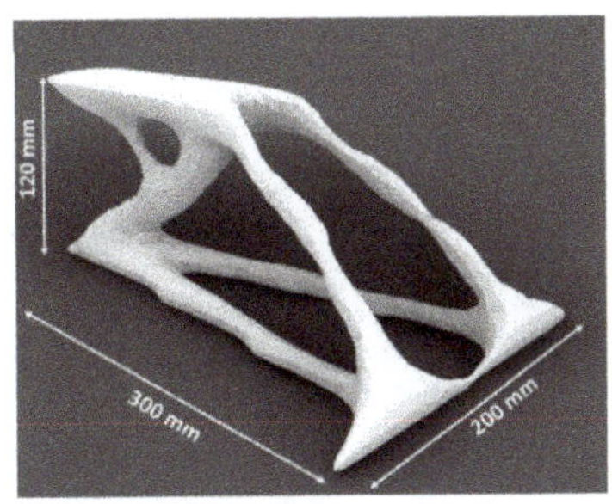

(b)

Figure 3. (**a**) Bounding box with constraints; (**b**) mesh after topology optimization.

Table 2. Mechanical properties of 316L stainless steel after DED process.

Property	Meltio XY Properties	Meltio XZ Properties
Yield Strength [MPa]	429 ± 6	347 ± 11
Ultimate Tensile Strength [MPa]	643 ± 16	655 ± 28
Elongation [%]	38 ± 2	41 ± 4

3. Results

In this section, the results from each step of the proposed workflow are presented.

3.1. Topology Optimization Bounding Box and Constraints

Starting with the topology optimization step, the resulting mesh that is extracted is shown in Figure 3b.

3.2. Orientation Optimization

Then, the mesh is rotated in every possible angle combination, aiming to find the one with the minimum support volume required. The optimal orientation that occurs from the orientation optimization process is shown in Figure 4a. The mesh is rotated by 180° around the X-axis and by 82° around the Y-axis with respect to the starting orientation.

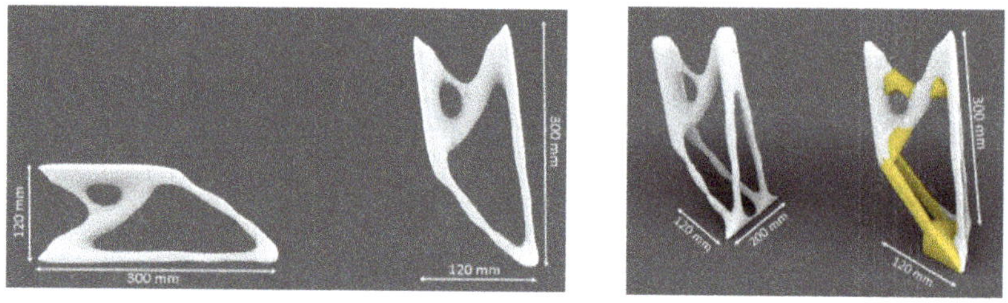

(**a**) (**b**)

Figure 4. (**a**) Mesh before (**left**) and after (**right**) orientation optimization; (**b**) mesh before (**left**) and after (**right**) material addition.

3.3. Material Addition

The next step is that the remaining overhanging areas are concealed with the suitable geometries generated from the material addition process described in the previous chapter. Figure 4b shows the fully optimized mesh next to the one that occurred from the orientation optimization procedure.

3.4. Slicing and Machine Code Generation

As the last step, the fully optimized mesh is sliced (Figure 5a), and the machine code to begin the manufacturing is exported. In this particular case, G-code is used; however, the method would be the same regardless of the specific machine programming language.

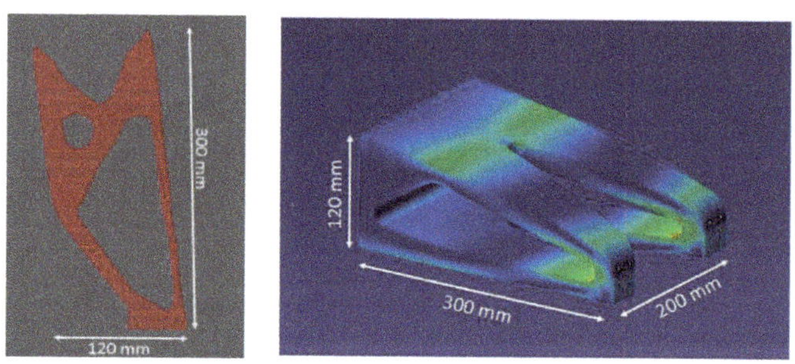

(**a**) (**b**)

Figure 5. (**a**) Contoured mesh (front view); (**b**) conventionally designed component.

Following the description of the proposed method to improve the manufacturability of a component through design optimization is the analysis of its effectiveness and evaluation of its results. In order to measure the effectiveness of the recommended workflow, a comparison with a conventional workflow of designing the same component is made. The parameters that will determine the evaluation procedure of the two workflows are the volume of the final meshes, the support volume of the final mesh, and the manufacturability time for a specific AM process. The values of the volumes are measured using Grasshopper 3D, while the manufacturing time is calculated using a commercial CAM software for AM (Ultimaker Cura), based on the specifications of a typical DED three-axis machine (Table 1).

The conventional workflow starts by importing the boundary domain in CAD software (in this case Dassault Systemes CATIA), and after applying the load case and the support conditions, a structural simulation is executed. The results of the simulation are then utilized to manually remove material from the initial domain in an iterative manner until the final component occurs (Figure 5b). The different colors on the part represent the different values of stress that are applied concerning the load and support scenario mentioned above. A blue color corresponds to zero value of stress, and as the stress value increases, the color changes to green, then yellow, and at last red, which represents the highest stress on the part.

The results of the conventional workflow and each sub-process of the proposed workflow concerning the aforementioned parameters are presented in Table 3.

Table 3. Comparison table between the parameters of the two workflows.

	Conventional Workflow	Proposed Workflow		
Parameter		**Step 1: Topology Optimization**	**Step 2: Orientation Optimization**	**Step 3: Material Addition**
Mesh volume [mm^3]	2667	507	507	723
Support volume [mm^3]	2723	654	103	40
Total volume [mm^3]	5390	1161	610	763
Build time [min]	13	6	10	9

4. Discussion

Taking into consideration the presented results from the conventional workflow and those that occur from the material addition sub-process, we can clearly come to the conclusion that the proposed method offers significant advantages regarding its effectiveness compared to the conventional one.

Regarding the results deriving from the conventional workflow and those from the topology optimization process, the difference is obvious. A total volume and build time reduction of around 80% and 50% is observed, respectively. Furthermore, it has to be noted that the conventional workflow includes an iterative process of manually (re)designing and simulating in order to minimize the part's volume, a procedure which is highly dependent on the designer experience, both in terms of the performance of the obtained part as well as the time required to design the part. On the other hand, the topology optimization process is completely automated, the results are consistent, and the whole workflow requires only a few minutes of runtime on an average PC.

In addition, it is evident that utilizing the proposed method significantly reduces the required support volume. This leads to a significant reduction in material usage but also in build time, as supports do not have to be deposited. What is not clearly presented in the above table, but is of equal significance, is the time savings that can be obtained due to the minimization or even elimination of the required subtractive post-processing towards removing support structures, which the proposed method eliminates. This results in a further reduction in the total time required to produce the parts, but also has direct cost reduction implications due to the elimination of cutting tools as well as of an additional process step (no setup time and no additional equipment needed).

The software tools used to implement the proposed method are Rhinoceros 3D and Grasshopper 3D due to their capability to combine a visualization environment and a simple visual programming language which also allows the user to embed their own custom code if needed (using Python or C#), combined with their affordability. The implementation of the proposed method can also be developed using other software tools, while the use of a programming language that includes 3D geometry libraries is proposed.

A limitation of the present work is the usage of bulk material properties in the topology optimization step. To obtain more realistic results, material coupons from the DED process to be used need to be characterized. In addition, actual manufacturing trials of the optimized component need to be completed, and the performance of the AM optimized part needs to be compared to the predicted values. However, this was beyond the scope of this particular work, which focuses on the development of a method that tackles manufacturability limitations by enhancing the design of a component that occurred from the topology optimization process, and not the verification of the topology optimization itself. Nevertheless, the authors intend to carry out such validation tests as a part of future work.

5. Conclusions

In this work, a process that optimizes the design of a given component to enhance its manufacturability for a specific AM procedure was presented. This workflow, in order to be accomplished, consists of certain steps which are the topology optimization, the orientation optimization, the material addition, and, to complete this process, a step that slices and generates machine code is also added. The results of this workflow regarding the mesh and the support structures' volume, as long as the estimated build time is compared to that of a conventional workflow, where the design engineer optimizes the component manually, are given as the outcome of a structural analysis. For the specific case study examined in the manuscript, a total volume and build time reduction of around 80% and 50% is observed, respectively, compared with a conventional workflow.

For the proposed method and tool, we have a direct practical implementation in mind, and they could find application in the automotive, aerospace, and architectural industries, where the need for optimized organic designs is constantly increasing, while optimizing the manufacturability is a key factor concerning decision making for production.

Future work will include validation tests to obtain realistic material properties, as well as comparing the predicted and achieved properties of the part through an application of this method in real parts. In addition, the proposed tool is to be improved and developed further so that it can produce a variety of components and geometries and be able to take into account additional AM manufacturability restrictions. Another goal is to upgrade the slicing and machine code generation sub-process to accommodate AM processes that can take advantage of five degrees of freedom (DoF) deposition.

Author Contributions: Conceptualization, P.S. and H.B.; methodology, H.B. and M.A.T.; software, M.A.T.; validation, M.A.T.; writing—original draft preparation, M.A.T.; writing—review and editing, H.B. and P.S.; supervision, P.S. All authors have read and agreed to the published version of the manuscript.

Funding: This research was co-funded by the European Regional Development Fund of the European Union and Greek national funds through the Operational Program Competitiveness, Entrepreneurship, and Innovation, under the call RESEARCH—CREATE—INNOVATE (project code: T2EDK-03896).

Data Availability Statement: Data sharing not applicable.

Conflicts of Interest: The authors declare no conflict of interest.

References

1. Stavropoulos, P. Introduction. In *Additive Manufacturing: Design, Processes and Applications*; SpringerBriefs in Applied Sciences and Technology; Springer: Cham, Switzerland, 2023. [CrossRef]
2. Attaran, M. The rise of 3-D printing: The advantages of additive manufacturing over traditional manufacturing. *Bus. Horiz.* **2017**, *60*, 677–688. [CrossRef]
3. Bikas, H.; Stavropoulos, P.; Chryssolouris, G. Additive manufacturing methods and modelling approaches: A critical review. *Int. J. Adv. Manuf. Technol.* **2015**, *83*, 389–405. [CrossRef]
4. Izdebska-Podsiadły, J. (Ed.) Chapter 3—Classification of 3D Printing Methods. In *Plastics Design Library, Polymers for 3D Printing*; William Andrew Publishing: Norwich, NY, USA, 2022; pp. 23–34, ISBN 9780128183113. [CrossRef]
5. Ahn, D.G. Directed Energy Deposition (DED) Process: State of the Art. *Int. J. Precis. Eng. Manuf.-Green Tech.* **2021**, *8*, 703–742. [CrossRef]
6. Grinde, S. Topology Optimization for Additive Manufacturing Using SIMP Method. Master's Thesis, Montana Technological University, Butte, MT, USA, 2018.
7. Bendsoe, M.P.; Kikuchi, N. Generating optimal topologies in structural design using a homogenization method. *Comput. Methods Appl. Mech. Eng.* **1998**, *71*, 197–224. [CrossRef]
8. MBendsoe, P.; Haber, R. The Michell layout problem as a low volume fraction limit of the homogenization method for topology design: An asymoptotic study. *Struct. Optim.* **1993**, *6*, 63–267. [CrossRef]
9. Bendsøe, M.P.; Sigmund, O. *Topology Optimization*; Springer: Berlin/Heidelberg, Germany, 2002.
10. Rozvany, G.I.N. *Structural Design via Optimality Criteria*; Kluwer: Dordrecht, The Netherlands, 1988.
11. Suzuki, K.; Kikuchi, N. A homogenization method for shape and topology optimization. *Comput. Methods Appl. Mech. Eng.* **1991**, *93*, 291–381. [CrossRef]

12. Allaire, G.; Kohn, R.V. Topology Optimization and Optimal Shape Design Using Homogenization. In *Topology Design of Structures*; Bendsoe, M.P., Soares, C.A.M., Eds.; NATO ASI Series, Series E; Kluwer: Dordrecht, The Netherlands, 1993; Volume 227, pp. 207–218. [CrossRef]
13. Allaire, G. The Homogenization Method for Topology and Shape Optimization. In *Topology Optimization in Structural Mechanics CISM*; Rozvany, G.I.N., Ed.; Springer: Berlin/Heidelberg, Germany, 1997; pp. 101–133. [CrossRef]
14. Rozvany, G.I.N.; Zhou, M.; Birker, T. Generalized shape optimization without homogenization. *Struct. Optim.* **1992**, *4*, 250–252. [CrossRef]
15. Bendsoe, M.P. Optimal shape design as a material distribution problem. *Struct. Optim.* **1989**, *1*, 193–202. [CrossRef]
16. Mjejnek, H.P. Some aspects of the genesis of structures. *Struct. Optim.* **1992**, *5*, 64–69. [CrossRef]
17. Bendsoe, M.P.; Sigmund, O. Material interpolations in topology optimization. *Arch. Appl. Mech.* **1999**, *69*, 635–654.
18. Rozvany, G. The SIMP Method in Topology Optimization—Theoretical Background, Advantages and New Applications. In Proceedings of the 8th Symposium on Multidisciplinary Analysis and Optimization, Long Beach, CA, USA, 6–8 September 2000. [CrossRef]
19. YXie, M.; Steven, G.P. A simple evolutionary procedure for structural optimization. *Comput. Struct.* **1993**, *49*, 885–896.
20. Xie, Y.M.; Steven, G.P. Optimal design of multiple load case structures using an evolutionary procedure. *Eng. Comput.* **1994**, *11*, 295–302. [CrossRef]
21. Chu, D.N.; Xie, Y.M.; Hira, A.; Steven, G.P. On various aspects of evolutionary structural optimization for problems with stiffness constraints. *Finite Elem. Anal. Des.* **1997**, *24*, 197–212. Available online: https://www.sciencedirect.com/science/article/pii/S0168874X96000492?via%3Dihub (accessed on 7 June 2023).
22. Xie, Y.M.; Steven, G.P. Evolutionary structural optimization for dynamic problems. *Comput. Struct.* **1996**, *58*, 1067–1073. Available online: https://www.sciencedirect.com/science/article/pii/0045794995002359 (accessed on 7 June 2023). [CrossRef]
23. Yang, X.Y.; Xie, Y.M.; Steven, G.P.; Querin, O.M. Bidirectional evolutionary method for stiffness optimization. *AIAA J.* **1999**, *37*, 1483–1488. [CrossRef]
24. David, M.A. *Design for Manufacturability; How to Use Concurrent Engineering to Rapidly Develop Low-Cost, High-Quality Products for Lean Production*; Routledge: Oxfordshire, UK, 2014; p. 486. Available online: https://www.design4manufacturability.com/books.htm (accessed on 7 June 2023).
25. Bikas, H.; Lianos, A.K.; Stavropoulos, P. A design framework for additive manufacturing. *Int. J. Adv. Manuf. Technol.* **2019**, *103*, 3769–3783. [CrossRef]
26. Zhang, Y.; Bernard, A.; Gupta, R.K.; Harik, R. Evaluating the design for Additive Manufacturing: A process planning perspective. *Procedia CIRP* **2014**, *21*, 144–150. [CrossRef]
27. Booth, J.W.; Alperovich, J.; Chawla, P.; Ma, J.; Reid, T.N.; Ramani, K. The Design for Additive Manufacturing Worksheet. *ASME. J. Mech. Des.* **2017**, *139*, 100904. [CrossRef]
28. Ghiasian, S.E.; Jaiswal, P.; Rai, R.; Lewis, K. From Conventional to Additive Manufacturing: Determining Component Fabrication Feasibility. In Proceedings of the ASME 2018 International Design Engineering Technical Conferences and Computers and Information in Engineering Conference, Quebec City, QB, Canada, 26–29 August 2018; Volume 2A. [CrossRef]
29. Lianos, A.; Koutsoukos, S.; Bikas, H.; Stavropoulos, P. Manufacturability Assessment and Design for AM. *Procedia CIRP* **2020**, *91*, 290–294. [CrossRef]
30. Borgue, O.; Müller, J.; Leicht, A.; Panarotto, M.; Isaksson, O. Constraint Replacement-based design for additive manufacturing of satellite components: Ensuring design manufacturability through tailored test artefacts. *Aerospace* **2019**, *6*, 124. [CrossRef]
31. Dinda, S. Designing a Mobile Application for Evaluating Manufacturability of Parts Using Additive Manufacturing. Master's Thesis, The Pennsylvania State University, State College, PA, USA, 2017. Available online: https://etda.libraries.psu.edu/catalog/14468sjd5520 (accessed on 7 June 2023).
32. Lianos, A.; Bikas, H.; Stavropoulos, P. A Shape Optimization Method for Part Design Derived from the Buildability Restrictions of the Directed Energy Deposition Additive Manufacturing Process. *Designs* **2020**, *4*, 19. Available online: https://www.mdpi.com/2411-9660/4/3/19 (accessed on 7 June 2023). [CrossRef]
33. O'Hara, R.P.; Chand, A.; Vidiyala, S.; Arechavala, S.M.; Mitsouras, D.; Rudin, S.; Ionita, C.N. Advanced 3D Mesh Manipulation in Stereolithographic Files and Post-Print Processing for the Manufacturing of Patient-Specific Vascular Flow Phantoms. *Proc. SPIE-Int. Soc. Opt. Eng.* **2016**, *9789*, 978909. [CrossRef]
34. Wang, Y.; Liu, Y.; Zou, J.; Zheng, J.; Yang, S. A novel two-phase evolutionary algorithm for solving constrained multi-objective optimization problems. *Swarm Evol. Comput.* **2022**, *75*, 101166. [CrossRef]
35. Niu, Y.; Shao, J.; Xiao, J.; Song, W.; Cao, Z. Multi-objective evolutionary algorithm based on RBF network for solving the stochastic vehicle routing problem. *Inf. Sci.* **2022**, *609*, 387–410. [CrossRef]
36. Wang, M.; Zhang, H.; Hu, Q.; Liu, D.; Lammer, H. Research and implementation of a non-supporting 3D printing method based on 5-axis dynamic slice algorithm. *Robot. Comput. -Integr. Manuf.* **2019**, *57*, 496–505. [CrossRef]
37. Brown, A.C.; de Beer, D. *Development of a Stereolithography (STL) Slicing and G-Code Generation Algorithm for an Entry Level 3-D Printer*; 2013 Africon, Pointe aux Piments, Mauritius; IEEE: Piscataway, NJ, USA, 2013; pp. 1–5. [CrossRef]
38. Kwame, S.; Dzegblor, N.K.; Lodonu, J.C. The use of computer-based tutorial to augment teaching and learning of computer software application: A case study of Rhinoceros 3d software. *Int. J. Innov. Res. Dev.* **2015**, *4*, 209–216. Available online: http://www.internationaljournalcorner.com/index.php/ijird_ojs/article/view/135425 (accessed on 7 June 2023).

39. Rhinoceros 3D Software. Available online: https://www.rhino3d.com/ (accessed on 9 June 2023).
40. Grasshopper 3D Software. Available online: https://www.rhino3d.com/features/#grasshopper (accessed on 9 June 2023).
41. Perini, K.; Chokhachian, A.; Dong, S.; Auer, T. Modeling and simulating urban outdoor comfort: Coupling Envi-met and TRNSYS by grasshopper. *Energy Build.* **2017**, *152*, 373–384. [CrossRef]
42. Mesh Normals. Available online: https://adamsturge.github.io/Engine-Blog/mydoc_mesh_normals.html (accessed on 9 June 2023).
43. van de Ven, E.; Maas, R.; Ayas, C.; Langelaar, M.; van Keulen, F. Continuous front propagation-based overhang control for topology optimization with additive manufacturing. *Struct. Multidiscip. Optimization.* **2018**, *57*, 2075–2091. [CrossRef]
44. Galapagos Plug-in for Grasshopper 3D. Available online: http://wiki.bk.tudelft.nl/toi-pedia/Galapagos_Optimization (accessed on 9 June 2023).
45. Dienemann, R.; Schumacher, A.; Fiebig, S. Topology Optimization Considering the Requirements of Deep-Drawn Sheet Metals. In Proceedings of the 11th World Congress on Structural and Multidisciplinary Optimization, Sydney, Australia, 7–12 June 2015.
46. Meltio Stainless Steel 316L Material Datasheet. Available online: https://meltio3d.com/wp-content/uploads/2023/04/Meltio_Stainless_Steel_316L.pdf (accessed on 22 August 2023).

machines

MDPI

Article

An Internet-of-Things-Based Dynamic Scheduling Optimization Method for Unreliable Flexible Manufacturing Systems under Complex Operational Conditions

Abdulmajeed Dabwan [1], Husam Kaid [1,*], Abdulrahman Al-Ahmari [2,3,*], Khaled N. Alqahtani [1] and Wadea Ameen [4]

1 Industrial Engineering Department, College of Engineering, Taibah University, Medina 41411, Saudi Arabia; adabwan@taibahu.edu.sa (A.D.); knqahtani@taibahu.edu.sa (K.N.A.)
2 Industrial Engineering Department, College of Engineering, King Saud University, Riyadh 11421, Saudi Arabia
3 Raytheon Chair for Systems Engineering (RCSE Chair), Advanced Manufacturing Institute, King Saud University, Riyadh 11421, Saudi Arabia
4 Industrial Engineering Department, College of Engineering and Architecture, Alyamamah University, Riyadh 11512, Saudi Arabia; w_qaid@yu.edu.sa
* Correspondence: hkaid@taibahu.edu.sa (H.K.); alahmari@ksu.edu.sa (A.A.-A.)

Abstract: The dynamic scheduling problem (DSP) in unreliable flexible manufacturing systems (UFMSs) with concurrency, conflicts, resource sharing, and sequential operations is a complex optimization problem that requires the use of efficient solution methodologies. The effectiveness of scheduling UFMSs relies on the quality of equipment maintenance. Currently, UFMSs with consistently large queues of parts awaiting service employ a repair-after-failure approach as a standard maintenance procedure. This method may require unexpected resources, incur costs, consume time, and potentially disrupt the operations of other UFMSs, either partially or fully. This study suggests using a predictive maintenance (PdM) strategy that utilizes the Internet of Things (IoT) to predict and avoid early mechanical equipment failures before they happen in UFMSs, thereby reducing unplanned downtime and enhancing reliability. Therefore, the objective of this paper is to construct timed Petri net (TPN) models using the IoT for the PdM configuration of mechanical equipment in the dynamic scheduling problem of UFMSs. This necessitates that users represent the specific problem using TPNs. The process of PN modeling requires the utilization of domain knowledge pertaining to the target problems as well as to machine information. However, it is important to note that the modeling rules for PNs are straightforward and limited in number. Consequently, the TPN model is applied to generate and formulate mixed-integer linear programming (MILP) instances accurately. This is done to identify the optimal production cycle time, which may be implemented in real-life scenarios. Several UFMS instances are used to demonstrate the applications and effectiveness of the proposed method. The computational results demonstrate that the proposed method shows superior solution quality, effectively solves instances for a total of 10 parts and 6 machines, and achieves a solution in a reasonable CPU time.

Keywords: flexible manufacturing systems; Petri net; Internet of Things; predictive maintenance; reliability; scheduling

1. Introduction

Flexible manufacturing systems (FMSs) are computer-controlled systems that can automatically execute a variety of activities based on predetermined process plans. The FMS includes a restricted number of resources, including machines, robots, automated guided vehicles (AGVs), and buffers, which are shared by multiple production processes within the system [1]. In an FMS, raw components are simultaneously processed in a predefined sequence to efficiently use constrained system resources and enhance overall system performance [2].

There are two major issues involved with the operation of an FMS: structural (design) issues and operational issues. The number of machine tools of each kind, the size of the material handling system, and the capacity of the buffers are FMS design considerations. FMS operational issues involve problems with planning, scheduling, and control. The decision of which parts should be sequentially machined and the assignment of pallets and fixtures to part kinds are planning issues. Given the selected part mix, the FMS scheduling problem involves deciding the optimal input order for components and the optimal order at each machine. The FMS control problems include observing the system to ensure that requirements and deadlines are met as well as addressing deadlock problems [3]. Real-world FMSs can be subject to unexpected uncertainties, including operator errors, machine malfunctions, and uncertain part processing times. Subsequently, the FMS encounters difficulties dealing with modifications in the manufacturing plans or orders. Hence, in order to enhance the promptness and adaptability of FMSs in dealing with uncertainties, it is imperative to optimize production scheduling, which is referred to as the dynamic scheduling problem (DSP). This entails adjusting the production scheduling strategies of FMSs based on the current production status in order to effectively adapt to the dynamic environment. Several combinatorial optimization problems (COPs) are NP-hard, and no solutions to these problems have been presented [4]. In the fields of manufacturing, agriculture, transportation, medical services, and sciences, combinatorial optimization can be used to minimize costs or maximize the impact of different systems. Optimization plays a crucial role in obtaining sustainable development objectives [5,6]. The era of big data has increased the significance of optimization studies, which extract important information from data using optimization methods [7–9]. Thus, this study has been considered one of the most significant subjects. If the feasible area is bounded, "integer programming" (IP) can be used to formulate combinatorial optimization problems [10]. This method is known as "integer linear programming" (ILP) if the objective function and all constraints of the COP are stated as linear equations. By combining integer and real variables in the mixed formulation, a greater number of COPs can be generated in comparison with simple IPs [11]; this is known as "mixed-integer programming" (MIP). When only linear constraints and objective functions are involved, therefore, MIP problems are referred to as "mixed-integer linear programming" (MILP) problems [12]. The FMS scheduling problem is a complicated optimization problem that is defined as NP-hard [13–16]. It is not feasible to find an exact solution to an FMS scheduling problem within an acceptable amount of time, even for instances of small size [16]. The optimal solution necessitates an exponential time frame. Hence, the solution to this NP-hard problem needs the utilization of metaheuristic methods. Researchers in the field of FMS scheduling have proposed several metaheuristic approaches, including the genetic algorithm (GA) [14–18], the ant colony optimization algorithm (ACO) [18,19], particle swarm optimization (PSO) [18,20], simulated annealing (SA) [21,22], Tabu Search (TS) [23,24], etc.

At present, simulation software is widely used for scheduling in several industries to guarantee the effective running of manufacturing processes, both technically and economically. Gholami and Zandieh [25] integrate the simulation and the genetic algorithm to minimize makespan and mean tardiness in flexible job-shop scheduling with machine stochastic breakdowns. Pergher et al. [26] combine simulation with the flexible and interactive tradeoff compensatory approach in job-shop production systems to determine the best combination of due order release, date assignment, and shop dispatching rules. Different combinations are assessed based on production quantity, total cost, total throughput time, and tardiness. Thenarasu et al. [27] integrate the simulation and the multi-criteria decision-making approaches to the model and evaluate flow time, makespan, and tardiness-based measures in partial flexible job-shop scheduling with both static and dynamic job arrivals. The open-source program Lekin is used to analyze and simulate the impact of dispatching rules on the makespan time in the flexible job-shop system [28]. The software also identifies the most efficient manufacturing process layout that minimizes production time [29].

Petri nets (PNs) are effective modeling and simulation tools for LP, ILP, and MILP problems because PNs contain mathematical expressions indicated as linear equations [30,31]. The work in [32] presented a method for constructing MILP based on PN models for combinatorial optimization problems involving "traveling salesman problems" and a "simple resource assignment problem". Tuncel and Bayhan [33] provided a comprehensive study of scheduling challenges in which they highlighted the integration of Petri nets with other approaches and addressed both theoretical developments and practical experiences. Existing approaches have been divided into four categories: (1) Petri net-based simulations connected with "heuristic dispatching rules" for the control and scheduling of manufacturing systems; (2) generative scheduling methods in which Petri nets are applied to develop schedules in terms of the transition firing sequences through the "reachability graph"; (3) a special Petri net class model employed to construct the planning and scheduling problem as a mathematical model; and (4) Petri net modeling approach providing a basis for the search procedure of metaheuristic methods to find the near-optimal "resource allocation" and the "event-driven schedule" in terms of the Petri net model firing sequences of the transitions. Wu et al. [34,35] constructed a Petri net model to represent wafer production processes in cluster tools requiring wafer revisitation. They proposed a systematic approach for evaluating their effectiveness, which leads to the derivation of optimality requirements for three-wafer period scheduling. Qiao et al. [36,37] constructed a model to represent wafer production processes using a "timed Petri net" to optimize their schedule. Analytical expressions can be used to determine whether the systems are schedulable. They also provided a simple implementation method for the obtained schedule. Using PNs, Zhou et al. [38] modeled and evaluated an FMS cell. They used "top-down refinement", "system decomposition", and "modular composition ideas" to obtain the structure and preservation of essential system characteristics, such as "liveness", "boundedness/safety", and "reversibility", which ensure the system's "stability", "deadlock-free", and "cyclic manner". A reduction approach was applied to transform the timed PN to an equivalent time-marked graph. Then, the typical method for determining cycle time for the marked graphs was implemented. In addition, Zhou et al. [39] developed PN models for FMS by scheduling the modeled FMS based on the firing sequences of the PN model from the initial marking to the final one. Using a class of timed PNs, they applied a branch-and-bound approach to determine the optimal schedule for the FMS. Artigues and Roubellat [40] proposed a PN model in the context of a generic approach for "on-line" scheduling in a job shop environment with various setup times and resources based on a detailed definition of essential states, decisions, and events for "on-line" and "off-line" scheduling. They used an "acyclic directed graph" to illustrate a set of static scheduling problem solutions. Mejia and Odrey [41] introduced Beam A* Search, a PN-based algorithm (BAS). This approach systematically increases segments of a PN "reachability graph" in order to identify a schedule that is close to optimal. Their proposed technique was evaluated by applying it on various FMS scheduling problems. Zhang et al. [42] modeled the assembly processes of the "flexible assembly system" using timed Petri nets (TPNs). A scheduling model was developed for the FAS, and a dynamic programming method was used to determine a viable allocation of processes to machines and to optimize the time to completion for either a "single product" or a "batch of products". In their study, Kim et al. [43] introduced a new scheduling approach for production systems that relies on the TPN model and a "reactive fast graph search" approach. The main objectives of this method are to minimize the maximum completion time (makespan) and the overall tardiness. In order to accomplish the objectives, they proposed a new search algorithm that integrates the "RTA*" with a "rule-based supervisor". Lee and Lee [44] developed heuristic functions for the A* method based on T-timed PNs for the purpose of optimizing the FMS scheduling problem by reducing the makespan. In addition, they developed enhanced versions of these heuristic functions, which obtained an initial near-optimal solution faster. Wang and Wang [45] investigated the FMS scheduling problem based on a PN with the objective of minimizing the makespan. Combining a "dynamic search window" with a "best-first algorithm" and "backtracking search", they proposed a hybrid heuristic search method for the scheduling problem. Kammoun et al. [46] developed a mathematical model that uses the properties of decomposed TPNs to address the FMS scheduling problem. The model aims to determine the optimal firing sequence of TPN transitions and minimize the total processing time. Moreover, a genetic algorithm was presented to find effective solutions for large-scale scheduling problems. With the objective of finding an optimal transition sequence that minimizes firing time, the authors of [47] proposed a new "admissible heuristic function" for scheduling FMSs utilizing P-timed PNs. Using the structural symmetry of a PN model of an FMS to make a partial reachability graph reduces the state explosion problem as much as possible. The estimate function is applied to each generated marking to calculate the cost of firing the transition sequence. PNs' ability to describe multiple states concisely, capture priority relations and structural connections, and model "deadlocks", "conflicts", "buffer sizes", and "multi-resource constraints" are the primary advantages of implementing PNs in the FMS scheduling issue compared with alternative approaches. This will assist system analysts in modeling complicated, integrated scheduling issues [45]. Most researchers who studied the scheduling problem in industrial systems used the "reachability tree". It is well known that the size of a "reachability tree" increases exponentially as the size of a PN increases. This makes it rather challenging to evaluate PN models with a significant number of places and transitions.

Several organizations are depending on a single maintenance approach to guarantee the optimal efficiency of their resources. Regrettably, only a small number of individuals take into account the expenses incurred by ignoring the monitoring of asset deterioration during the early phases of defect formation. Maintenance can be categorized into three distinct types: corrective maintenance (CM), preventive maintenance (PM), and predictive maintenance (PdM). PM and CM primarily revolve around the age of the asset and adhering to a regular maintenance schedule. However, their drawback lies in the fact that they address resource repairs only when they have reached an advanced degree of deterioration. Frequently, these resources experience failures during the intervals between inspections, resulting in escalated repair expenses and significant hazards to human safety. Many companies mistakenly believe that a strategy that combines breakdown maintenance and the life of the resource is enough to accurately anticipate resource failure. Frequently, they neglect significant faults that result in severe malfunction and necessitate the replacement of resources rather than their repair, resulting in additional expenses. During CM, maintenance workers promptly commence work upon the occurrence of a problem. The objective of CM is to expedite the restoration of systems to their normal functioning state. CM does not involve a scheduled program for normal maintenance. Maintenance can take place only when there is an existing problem. The cost of repairs may be marginally higher, but it is significantly more affordable than the expense of routinely employing staff to maintain equipment. The equipment is repaired promptly; however, this approach can have adverse consequences in the event of a catastrophic occurrence. PdM is an improved approach to performing maintenance activities. It involves using test results and trends to anticipate or identify issues with a component of equipment. These methods employ noninvasive testing procedures to quantify and calculate trends in equipment performance. Some examples include vibration analysis, infrared analysis, thermography, ultrasound, and various other techniques.

The IoT enables the connection of physical things, enabling them to communicate information via the Internet. This capability has the potential to facilitate the collection of large volumes of data, which may be a strong asset for the success of businesses and future predictions [48,49]. At present in the field of maintenance, the use of embedded hardware consisting of sensors and other intelligent equipment controlled by the IoT is changing industrial and manufacturing operations [50,51]. The utilization of IoT in the predictive maintenance approach can be highly effective in this context. It involves directly monitoring equipment by continuously capturing real-time data on key stress-related variables, such as noise level, temperature, vibration, pressure, power consumption, and other interconnected devices [52]. This allows users to gain visibility into the performance of their assets and discover valuable insights. It also enables the detection of anomalies, the identification of patterns, and the recognition of warning signals that may indicate an imminent failure [52,53].

As mentioned in the literature, repairing UFMS equipment is a standard procedure because the equipment often fails without any prior communication to the maintenance team. As a result, the detection of faults and the subsequent unplanned maintenance can disrupt the current operations of the remaining UFMSs, either fully or partially. In order to enhance the use, availability, and reliability of equipment and minimize the costs associated with equipment maintenance, it is crucial for the maintenance team to have access to real-time condition-based data. This allows for efficient repairs and minimizes downtime, unnecessary inspections, maintenance time, and undue pressure on the maintenance team [54,55]. Motivated by the issues mentioned previously, this study identifies a lack of adoption of current PdM systems for this type of equipment. This is caused by variations in mechanical characteristics among various equipment and the absence of historical or real-time performance data. Therefore, the main contribution of this paper is to develop TPN models for the dynamic scheduling problem of UFMSs with concurrency, conflicts, resource sharing, and sequential operations. Subsequently, the proposed TPN model is utilized to generate the MILP in order to determine the optimal production cycle time, which can be implemented in real-life situations. Additionally, the paper proposes the utilization of the IoT for the PdM configuration to prevent early mechanical equipment malfunctions in UFMSs and avoid interruptions in the scheduled operations of other UFMSs. Finally, several cases of UFMSs are used to demonstrate the practical applications and effectiveness of the proposed approach.

The rest of the paper is organized as follows: Section 2 introduces timed Petri net synthesis and the estimation of resource failure time using the Internet of Things. Section 3 describes a statement and MILP model based on TPN to address the dynamic scheduling problem in unreliable flexible manufacturing systems with concurrency, conflicts, resource sharing, and sequential operations constraints. Section 4 demonstrates the practical use and effectiveness of the developed MILP model through the use of instances. Section 5 presents a brief summary and outlines the future research endeavors of the study.

2. Preliminaries

2.1. Timed Petri Nets Synthesis

Petri nets are a mathematical modeling tool that involves symbols such as place, transition, arc, and token, as depicted in Figure 1. The original PN did not indicate any time aspect. Timed Petri nets (TPNs) are created by including time in the PN to evaluate performance associated with time. TPNs are formally defined as a six-tuple $N = (P, T, F, W, h(T), M)$ where:

1. $P = P_A \cup P_R$: represents a non-empty finite set of model places, where $P_A = \cup_{i\in |m|}\{p_i\}$, $m > 0$ and $P_R = \cup_{i\in k}\{p_{mk}\}$, $k > 0$ represent sets of model operation and resource places, respectively;
2. $T = \cup_{j\in |T|}\{t_j\}$: represents a non-empty finite set of model transitions, $P \cap T = \emptyset$, and $P \cup T \neq \emptyset$;
3. $F \subseteq (P \times T) \cup (T \times P)$: represents flow relations and is denoted by arcs connecting places to transitions $(P \times T)$ or from transitions to places $(T \times P)$;
4. W: $(P \times T) \cup (T \times P) \rightarrow \mathbf{IN}$: is the weight-to-arc affectation function, where $\mathbf{IN} = \{0, 1, 2, \ldots\}$.
5. $h(t_j) \in h(T)$: denotes the static time that is assigned to t_j, $t_j \in T$;
6. ${}^\bullet p_i(p_i{}^\bullet)$: represents the input transitions (and corresponding output) of the place p_i;
7. ${}^\bullet t_j(t_j{}^\bullet)$: represents the input places (and corresponding output) of the transition t_j;

8. $M: P \rightarrow \mathbf{IN}$ denotes the function of the marking, which allocates to $p_i \in P$ a number of tokens, the initial marking is denoted by M_o, and place marking is represented by $M(p_i)$;
9. $M[t_j\rangle$: represents that at marking M, the transition t_j can be fired.

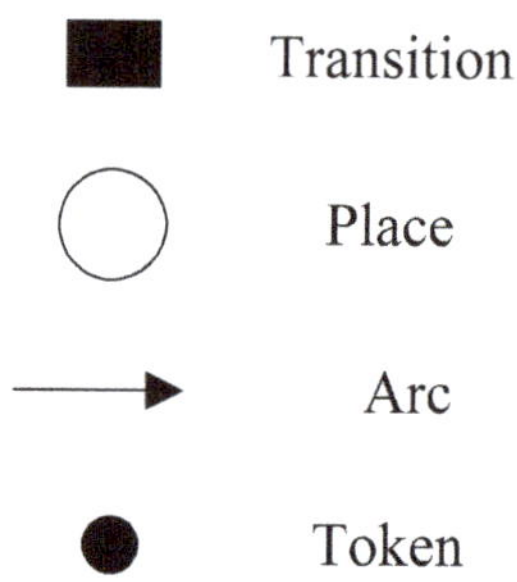

Figure 1. Icons for the Petri net symbols.

It is essential to note that marking M enables the transition t_j to be fired if $M(p_i) \geq W(p_i, t_j)$, $t_j \in p_i^\bullet$. When the transition $t_j \in T$ is enabled at marking M, the marking M can be modified from M to a new marking M', designated as $M[t\rangle\, M'$ and expressed as follows:

$$M'(p_i) = \begin{bmatrix} M(p_i) + W(p_i,\ t_j) & \text{if } p_i \in {}^\bullet t_j \setminus t_j^\bullet \\ M(p_i) - W(t_j, p_i) & \text{if } p_i \in t_j^\bullet \setminus {}^\bullet t_j \\ M(p_i) + W(t_j,\ p_i) - W(p_i,\ t_j) & \text{if } p_i \in t_j^\bullet \cap {}^\bullet t_j \\ M(p_i) & \text{otherwise} \end{bmatrix} \tag{1}$$

To represent the synthesis of the TPN model, consider Figure 2, which consists of three places and two transitions. The transition t_1 is enabled at the initial marking $M_o = p_1 + 0p_2 + 0p_3$, i.e., $M_o[t_1\rangle$. At M_o, the transition t_1 fires after two time slices (t.s). It selects a token from p_1 and deposits one token into p_2. Then, it creates a new marking $M_1 = 0p_1 + p_2 + 0p_3$ and expresses it as $M_o[t_1\rangle\, M_1$. Similarly, at M_1, the transition t_2 is enabled and can be fired with a delay of 4 t.s to create a succeeding marking, which can be expressed as $M_1[t_2\rangle\, M_2$ with $M_2 = p_1 + 0p_2 + p_3$.

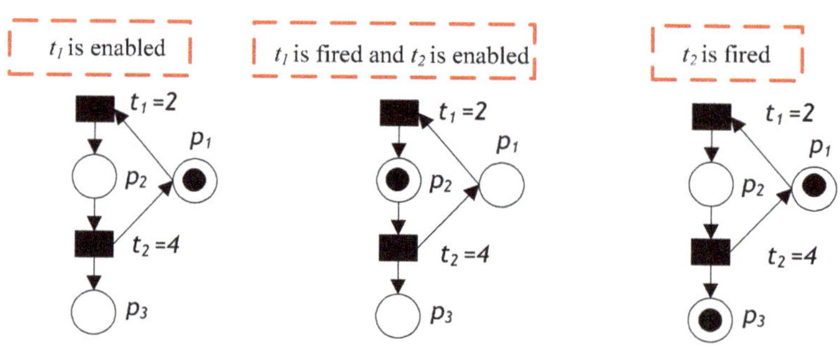

Figure 2. An example of a timed Petri net.

There are typically three types of places for PN models of FMSs: the set of idle places, the set of operation places, and the set of resource places. The number of tokens in the corresponding operation place at the initial marking indicates the maximum number of jobs that can be processed concurrently for a certain job type. In this study, an idle place is referred to as a sink place. The sink place is free of output transitions. A token at an initial operation place denotes a raw job that is ready for processing, while a token in a sink place indicates a finished job. In an FMS, resource places represent the resources (such as machines and robots). At the initial marking, the tokens in a resource place reflect the available resource units. The operation places, which are initially unmarked, represent the procedures that must be completed for the jobs in the production sequences. On the basis of timed PNs, each transition is associated with a time delay that indicates the time that is required to perform the appropriate operation. When the token's time delay finishes, it becomes available at an operation place.

In order to demonstrate the modeling of TPNs, consider an FMS, presented in Figure 3, that is composed of two machines m = 1, 2, operating two different jobs i = 1, 2, in a sequential manner; one robot, r_1 (material-handling equipment); and a load/unload station.

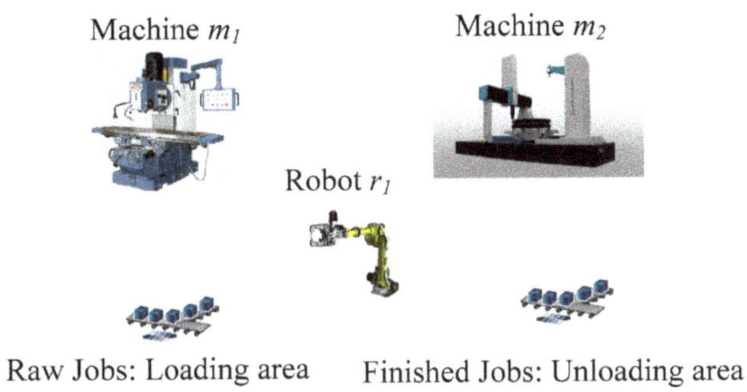

Figure 3. An example of an FMS.

Table 1 displays the sequence of job-production operations, the duration of each operation, and the machine that can be used for each operation. As shown in Table 1, each job involves a set of operations. For instance, job 1 contains two operations that sequentially require m_1 and m_2. Thus, job 1 can be completed sequentially using machines m_1 and m_2 with corresponding processing times of 4 and 6, respectively.

Table 1. Scheduling data of an FMS shown in Figure 3.

Job	Operation	Time	Machine
1	1	4	1
	2	6	2
2	1	8	2
	2	10	1

Figure 4 illustrates the TPN model for this system, and the description of places and transitions in the TPN model is given in Table 2.

Table 2. Description of places and transitions of the TPN model shown in Figure 4.

Places	Description
p_{11}	Process 1 for part 1
p_{12}	Process 2 for part 1
p_{21}	Process 1 for part 2
p_{22}	Process 2 for part 2
p_{m1}	Machine 1 is available
p_{m2}	Machine 2 is available
t_{11}	End process 1 of part 1
t_{12}	End process 2 of part 1
t_{21}	End process 1 of part 2
t_{22}	End process 2 of part 2

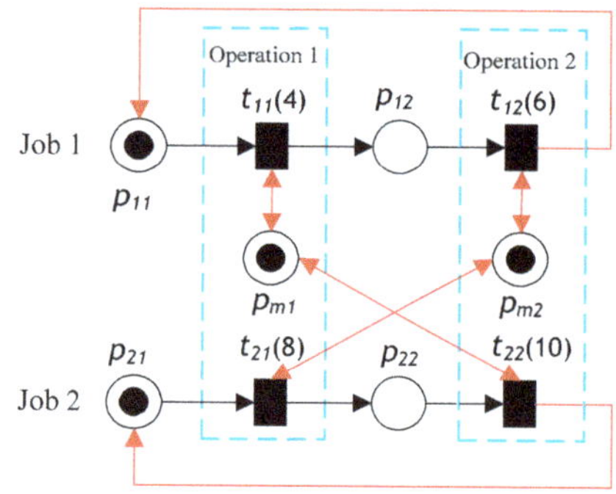

Figure 4. A timed PN model of the system shown in Figure 3.

2.2. Estimation of Resource Failure Time Using IoT

In the context of FMSs, the term "resource failure" denotes an issue characterized by temporal uncertainty. In case of a "resource failure", it is necessary to establish a "recovery subnet" with the ability to fix the problem [56]. The resource possesses the characteristic of reusability. This section aims to provide formal definitions for the purpose of estimating the occurrence of faults in an FMS using the IoT. It is assumed that the system contains sensors for the purpose of detecting the occurrence of resource breakdowns. The sensors facilitate the transmission of data to the Internet. Consequently, the gathered data can be accessed from any geographical point across the world. After the transmission of data to the Internet, a computer system will initiate the process of downloading and then utilize the acquired data to identify and rectify any cases of malfunction or failure. Hence, in the case of a failure of any given resource, online operations can persist by utilizing the availability of other resources. The value obtained from the sensors serves as a "set point" or "threshold value" to aid the system in determining its self-regulation strategy. Figure 5 illustrates the flowchart and architecture of the developed systems.

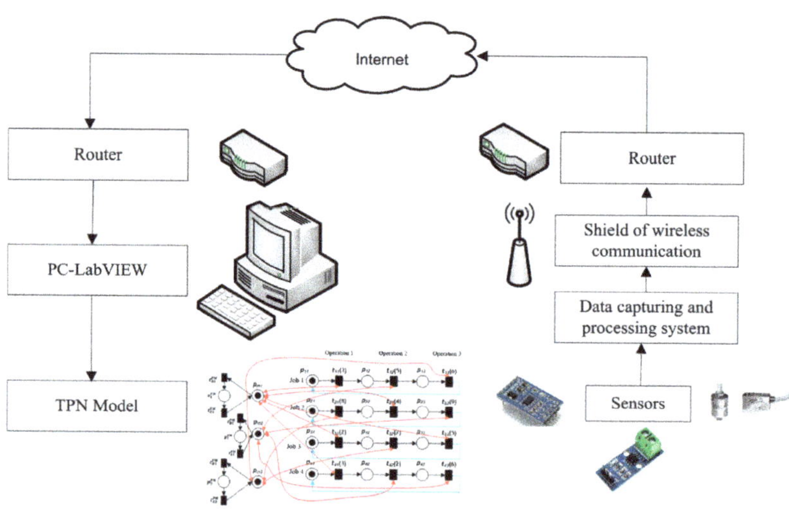

Figure 5. Resources failure detection flowchart.

Consider a TPN denoted by N, where N is defined as $N = (P, T, F, W, h(T), M_o)$. Let N_{IoT} be an estimation resource of the p_i failure time and recovery net based on IoT, denoted as $N_{IoT} = (\{p_{mi}, p_{Sij}, p_{DT}, p_{RT}, p_X, p_{PC}, p_{Rmi}\}, \{t_{Imi}, t_{fmi}, t_{tmi}, t_{DT}, t_{RT}, t_X, t_{PC}\}, F_{IoT})$, where M_{IoTo} indicates the initial markings of N_{IoT}, $M_{IoTo}(p_i) \geq 0$. $F_{IoT} = \{(t_{Imi}, p_{Sij}), (p_{Sij},$

t_{DT}), (t_{DT}, p_{DT}), (p_{DT}, t_{RT}), (t_{RT}, p_{RT}), (p_{RT}, t_X), (t_X, p_X), (p_X, t_{PC}), (t_{PC}, p_{PC}), (p_{PC}, t_{fmi}), (p_{mi}, t_{fmi}), (t_{fmi}, p_{Rmi}), (p_{Rmi}, t_{rmi}), (t_{rmi}, p_{mi})}, where p_{Sij} represents the sensor's system for resource p_i failure sensing, p_{DT} indicates the "data capture with wireless shield", and p_{RT} denotes the router. The data collected by the p_{Sij} sensors are transmitted to the Internet, specifically to the servers of Xively (expressed as p_X), which is a platform specifically developed for the IoT. The data that have been gathered can be accessed and observed through the PC-LabVIEW program, which is denoted as p_{PC}. The transitions t_{Imi}, t_{fmi}, t_{rmi}, t_{DT}, t_{RT}, t_X, and t_{PC} represent various types of transitions within the system. Specifically, these transitions correspond to the resource input sensors data p_{mi}, the resource failure time p_{mi}, the resource recovery of the resource p_{mi}, the data capture, the router, the Xively, and the PC/lab view, respectively. An unreliable net is formed when the TPN (N, M_o) is joined with the estimating resource failure net (N_{IoT}, M_{IoTo}), as shown in Figure 6. This combination is denoted by the formula (N_S, M_{So}) = (N_{IoT}, M_{IoTo}) || (N, M_o), where || indicates the combination of (N_{IoT}, M_{IoTo}) and (N, M_o).

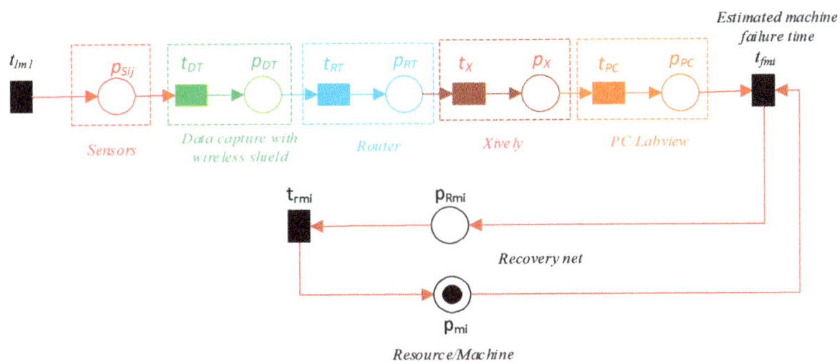

Figure 6. Estimation resource failure time using IoT.

Algorithm 1 illustrates the steps of synthesizing a network for estimating resource failure and recovery using the IoT. Algorithm 1 has the capability to rapidly identify changes in sensors through the monitoring function software integrated into the resource failure monitoring system. When an important malfunction happens in any machine within the system, there is a substantial increase in the amplitude of the sensors. The software will issue a notification in the event of a substantial breakdown in the resource and when the sensor data reach the predetermined threshold.

In order to demonstrate the synthesizing of an estimating resource failure time using the IoT, reconsider the TPN presented in Figure 4. It has one unreliable resource, which is p_{m1} (machine 1). The unreliable resource p_{m1}, as illustrated in Figure 7, consists of three sensors designed to identify tool failure: p_{s11} for the current sensor, p_{s13} for the accelerometer sensor, and p_{s13} for the acoustic emission sensor. The data transmission procedure to the Internet comprises multiple different phases.

During the initial phase, data readings from the sensors p_{s11}, p_{s12}, and p_{s13} are obtained (via t_{Imi}) and subsequently transmitted to the "Wi-Fi wireless shield" represented by t_{DT} and p_{DT}. Subsequently, the shield transmits the data to a wireless router, denoted by t_{RT} and p_{RT}. Integration of objects into a communication network constitutes a fundamental component of the IoT. Fundamentally, it represents an innovative methodology in which each entity is fully interconnected via the Internet infrastructure and participates in immediate transfers. In practical terms, the IoT refers to the seamless connection of equipment, sensors, and common household items via wired or wireless networks with the Internet. The implementation of this methodology is viable, and its operation will yield financial benefits due to the pervasive availability of the Internet. The integration of this technology will enable seamless integration of sensors across various automation applications, including residential and commercial settings. Therefore, a correlation between an object and a digital network can be established, facilitating the storage of the object's data and their subsequent visualization in the physical environment. This study presents a PC-LabVIEW-based system that has been designed to send Internet data to the TPN. The data stored in the cloud are retrieved and sent to the TPN at t_{fmi} for analysis and transmission. This is performed using a PC that is equipped with LabVIEW, referred to as the t_{PC} and p_{PC}. The TPN implements the OPC communication standard to implement the execution of the PC-LabVIEW program for estimating resource p_{m1} failure time.

Algorithm 1: *Synthesizing of estimating resource failure time and recovery net using the IoT.*

Input: *The TPN with* $N = (P, T, F, W, h(T), M_o)$;
Output: *The TPN based on IoT* (N_S, M_{So});

1. *To the given TPN*
   1.1 *Add places* $p_{DT}, p_{RT}, p_X, p_{PC}, p_{Rmi}$;
   1.2 *Insert transitions* $t_{DT}, t_{RT}, t_X, t_{PC}$;
   1.3 *Connect arcs* $(t_{DT}, p_{DT}), (p_{DT}, t_{RT}), (t_{RT}, p_{RT}), (p_{RT}, t_X), (t_X, p_X), (p_X, t_{PC}), (t_{PC}, p_{PC})$;
2. ***for*** $(1 \leq |P_R| \leq k{+}{+})$, ***do****/* for all unreliable places */*
   2.1 *Insert an input transition* t_{Imk} *of place* p_{mk};
   2.2 *Insert a failure transition* t_{fmk} *of place* p_{mk};
3. ***for*** $(1 \leq \psi_{ki} \leq i{+}{+})$, ***do****/** ψ_{ki} *is the number of sensors in* p_k**/*
   3.1 *Add a sensor place* p_{Skj} *to* p_{mk};
   3.2 *Connect an arc from* t_{Imk} *to* p_{mk};
   3.3 *Connect an arc from* p_{Skj} *to* t_{DT};
4. ***end for***
5. **Connect** *arcs* $(p_{PC}, t_{fmk}), (p_{mk}, t_{fmk}), (t_{fmk}, p_{Rmi}), (p_{Rmi}, p_{mk})$;
6. ***end for***
7. ***Output***: *The net* (N_S, M_{So});
8. ***End***

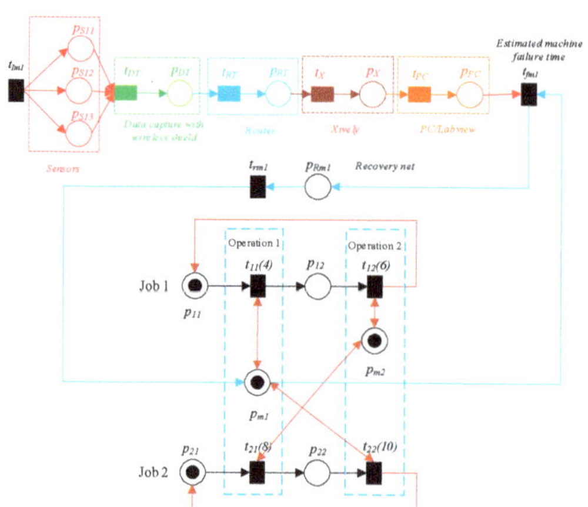

Figure 7. Estimating resource failure time and recovery net using the IoT for Figure 6.

3. Problem Formulation

3.1. Problem Definition

This study concentrates on a specific type of FMS where the machine's flexibility allows it to perform a variety of tasks. The machine's time taken to switch between modes of operation is insignificant. The studied FMS is described as a compilation of multiple job types with a predetermined and well-known processing sequence for all jobs inside the machine set. In the FMS scheduling problem, there are n independent job types, Ω independent machines, and Ψ_i operations. Indexes i,j represent job types, m denotes machines, and k,l indicate operations, where $i,j = 1, 2, \ldots, n$; $m = 1, 2, \ldots, \Omega$; and $k,l = 1, 2, \ldots, \Psi_i$. Each job $_i$ has a predetermined sequence of Ψ_i operations, and each operation has to be processed on a predetermined machine m denoted as $O_{i,k,m}$ with processing time $D_{i,k}$. The following assumptions are considered:

1. n jobs of various types are initially inserted in the initial place.
2. The starting and finishing places of each job type are represented, respectively, by p_{i1} and p_i, where $i = 1, 2, \ldots, n$.
3. At time zero, all machines are available for job processing.
4. All jobs are available for processing at time zero.
5. Each job is processed based on its defined and known sequence through the machine set.
6. Due to PdM actions, machines could be unavailable for a period of time.
7. The processing time varies by job type.
8. Effective part processing (no rework allowed).
9. Each machine can process only one type of job at a time.
10. Each job must be completed without interruption.

The purpose is to find the optimal sequence of parts and minimize the overall time it takes to complete them. Furthermore, our focus lies on implementing an efficient way to construct the production system supervisor that is connected to the first TPN model.

3.2. MILP Formulation

PN time constraints are typically related to places or transitions. In this study, TPNs are employed. The graphical models of TPNs are constructed from the logical sequence of the modeled system. Therefore, they are simple to build, extend, and apply to production system scheduling problems. In FMS, there are three major precedence relations for part processing: operations are processed in accordance with specific sequences, operations are processed in any sequence, or both relations are included in a part process plan. Synchronization and concurrency principles are essential to the study of FMS, and TPNs enable simple structures to describe these crucial concepts, as well as "conflicts", "non-determinism", "timing information", and "resource sharing environment". In this paper, the problem is FMS scheduling with PdM to minimize the total completion time.

3.2.1. Decision Variables

The FMS scheduling problem requires the assignment of machines to all jobs without machine conflicts. Such a machine assignment involves job scheduling. To represent the behavior of Petri nets, three types of decision variables are used. First, for all jobs, $t_{i,k}$ denotes the transition for the operation k of a job i; the firing start and end times of operation k for job i are represented as follows:

$$h\left(t_{i,k}^{Start}\right),\ h\left(t_{i,k}^{End}\right) \geq 0 \qquad \forall i,k \qquad (2)$$

Second, the machine m assignment for each operation k of a job i is defined, which provides the conditional information of the token for the transition $t_{i,k}$ to fire, as follows:

$$X_{i,k,m} \in \{0,1\} \qquad \forall i,k,m \qquad (3)$$

where $X_{i,k,m} = 1$ indicates that the machine m is allocated to a transition $t_{i,k}$, and $X_{i,k,m} = 0$ otherwise.

Third, to define the precedence relation of operations for a job i in the scheduled Petri net, the following variables are defined:

$$Y_{i,k,l} \in \{0,1\} \qquad \forall i,k,l \qquad (4)$$

where $Y_{i,k,l} = 1$ indicates that the operation l is processed after operation k for job i, and $Y_{i,k,l} = 0$ otherwise.

Fourth, to address the machine conflict between jobs i and j, the following variables are used to determine the utilization priority of shared machines:

$$O_{i,k,m,j,l} \in \{0,1\} \qquad \forall i,j,k,l,m \tag{5}$$

where $O_{i,k,m,j,l} = 1$ indicates that the operation l for job j is processed after operation k for job i in machine m, and $O_{i,k,m,j,l} = 0$ otherwise.

Finally, to represent the maintenance activity for machine m, the following variables are used:

$$Z_{i,k,m} \in \{0,1\} \qquad \forall i,k,m \tag{6}$$

where $Z_{i,k,m} = 1$ denotes that the maintenance activity is conducted before processing job i in machine m, and $Z_{i,k,m} = 0$ otherwise.

3.2.2. Constraints and Objective Function

The constraints and objective function for the FMS scheduling can be formulated by extracting the required information from the timed PN model. To execute an operation k for a job i, a machine m must be allocated to each operation defined by R_i, where R_i denotes $^\bullet t_{i,k} \cap P_R$, or the set of resource places associated with $t_{i,k}$. M_0 contains tokens representing the available machines. To allocate the necessary machines to each job i, the following constraints are used:

$$\sum_{i=1}^{\Omega} X_{i,k,m} \leq 1 \qquad \forall i,k. \tag{7}$$

As shown in Figure 8, the completion time of the operation k for each job i depends on the processing time of operation k and the time to repair for corrective maintenance of the assigned machine m and can be formulated as follows:

$$h\left(t_{i,k}^{End}\right) \geq h\left(t_{i,k}^{Start}\right) + D_{i,k} \cdot X_{i,k,m} \qquad \forall i,k,m \tag{8}$$

where $D_{i,k}$ is the processing time of operation k in the job i.

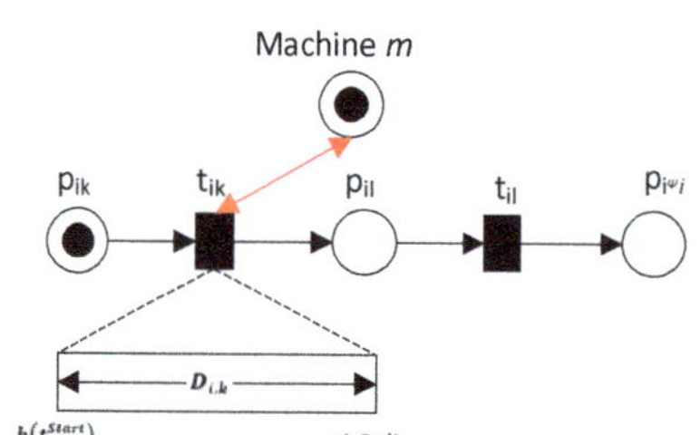

Figure 8. An example of the completion time of an operation for a job.

Figure 9 illustrates the sequence of operations when operation k precedes operation l for job i. The processing time of the first operation (k) is included in the firing time of the end transition $h\left(t_{i,k}^{End}\right)$. If there are two operations in series, as shown in Figure 8, then the time interval between the firing completion of both the end transitions $h\left(t_{i,k}^{End}\right)$ and $h\left(t_{i,l}^{End}\right)$ must be equal to or larger than the operation time of the second operation. By testing the condition $t_{i,k}{}^\bullet \cap P_A = {}^\bullet t_{i,l} \cap P_A$, the precedence relation can be obtained from the timed PN model shown in Figure 9, where $t_{i,k}{}^\bullet \cap P_A = {}^\bullet t_{i,l} \cap P_A = p_{il}$. This constraint can be represented as follows:

$$h\left(t_{i,l}^{End}\right) - h\left(t_{i,k}^{End}\right) \geq D_{i,l} \qquad \forall i,k,l\ t_{i,k}{}^\bullet \cap P_A = {}^\bullet t_{i,l} \cap P_A \tag{9}$$

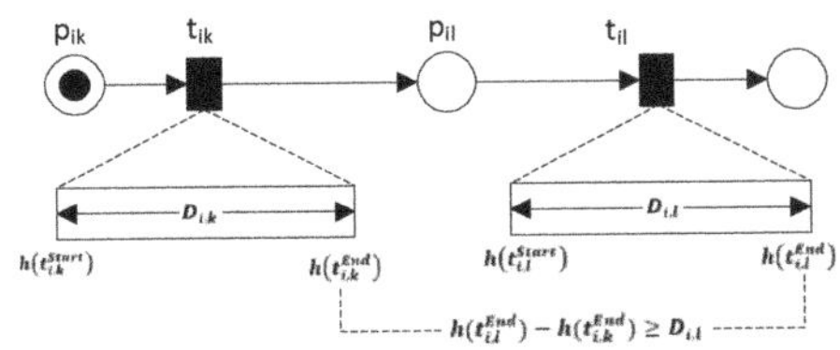

Figure 9. Start and end of operations for job i (operations k and l).

In many cases, some operations in a job are not linked by precedence. This can be modeled in TPN by utilizing the synchronization and mutual exclusion properties of PN, as shown in Figure 10. In this case, the machine can start processing the selected operation without precedence restriction, but both operations will be processed in a certain sequence because the job entity is considered a shared job.

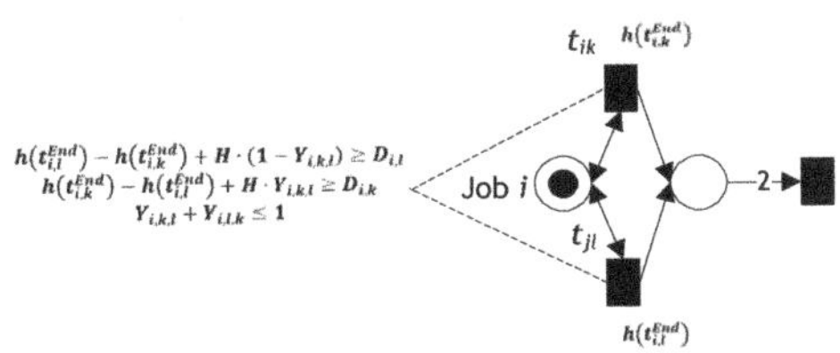

Figure 10. Representation of two independent operations for the same job type.

By checking the condition $t_{i,k}{}^{\bullet} \cap P_A = {}^{\bullet}t_{i,l} \cap P_A$, the precedence relation can be derived from the TPN model. The precedence variable, $Y_{i,k,l}$ is introduced in (4) to prevent machine conflicts, which can be formulated as

$$h\left(t_{i,l}^{End}\right) - h\left(t_{i,k}^{End}\right) + H\cdot(1 - Y_{i,k,l}) \geq D_{i,l} \qquad \forall i,k,l,k \neq l,\ t_{i,k}{}^{\bullet} \cap P_A = {}^{\bullet}t_{i,l} \cap P_A \quad (10)$$

$$h\left(t_{i,k}^{End}\right) - h\left(t_{i,l}^{End}\right) + H\cdot Y_{i,k,l} \geq D_{i,k} \qquad \forall i,k,l,k \neq l,\ t_{i,k}{}^{\bullet} \cap P_A = {}^{\bullet}t_{i,l} \cap P_A \quad (11)$$

$$Y_{i,k,l} + Y_{i,l,k} \leq 1 \qquad \forall i,k,l,k < l,\ t_{i,k}{}^{\bullet} \cap P_A = {}^{\bullet}t_{i,l} \cap P_A \quad (12)$$

where H represents a sufficiently large number.

As shown in Figure 11, if the same machine m is allocated to both jobs i and j, there cannot be any processing overlap. The precedence variable $O_{i,k,m,j,l}$ is introduced in (5), to prevent machine conflicts.

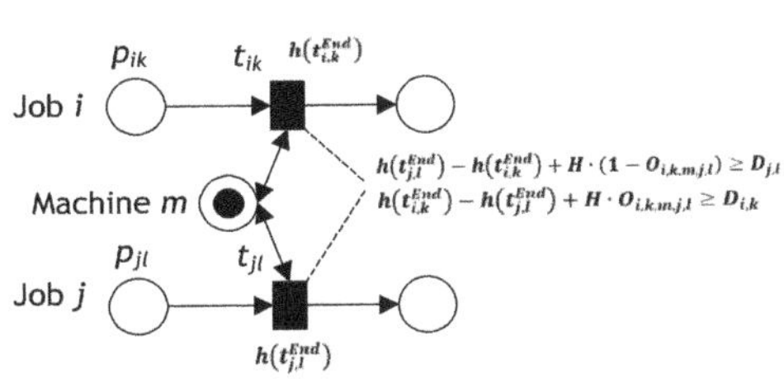

Figure 11. Shared machine in an FMS for jobs i and j.

If $O_{i,k,m,j,l} = 1$, Constraints (13) and (14) guarantee that job j cannot start before job i is completed. Moreover, Constraints (15) and (16) are required to guarantee that the variables $O_{i,k,m,j,l}$ have the correct values: $O_{i,k,m,j,l}$ or $O_{j,l,m,i,k}$ should be either one or zero if jobs i and j use the same machine m; otherwise, both variables must be zero.

$$h\left(t_{j,l}^{End}\right) - h\left(t_{i,k}^{End}\right) + H\cdot\left(1 - O_{i,k,m,j,l}\right) \geq D_{j,l} \qquad \forall i,j,k,l,i \neq j, R_i \cap R_j \neq \varnothing \tag{13}$$

$$h\left(t_{i,k}^{End}\right) - h\left(t_{j,l}^{End}\right) + H\cdot O_{i,k,m,j,l} \geq D_{i,k} \qquad \forall i,j,k,l,i \neq j, R_i \cap R_j \neq \varnothing \tag{14}$$

$$O_{i,k,m,j,l} + O_{j,l,m,i,k} \leq 1 \qquad \forall i,j,k,l,i < j, R_i \cap R_j \neq \varnothing \tag{15}$$

$$X_{i,k,m} + X_{j,l,m} - O_{i,k,m,j,l} - O_{j,l,m,i,k} \leq 1 \qquad \forall i,j,k,l,i < j, R_i \cap R_j \neq \varnothing \tag{16}$$

Constraint (17) guarantees that the end firing time of a transition for any operation is greater than the corresponding processing time.

$$h\left(t_{i,k}^{End}\right) \geq D_{i,k} \qquad \forall i,k \tag{17}$$

In Figure 10, the shared resource is the job, while in Figure 9, the shared resource is the machine. The proposed formulation restricts the use of shared resources to an optimal sequence. As shown in Figure 12, consider that all FMS machines are available at time 0. Each machine m is subject to failure, and it is assumed that the time to failure can be estimated using the IoT. To provide higher machine reliability, the age of a machine cannot exceed a maintenance period $[t_m^f, t_m^R]$, where t_m^f and t_m^R indicate the time to failure and repair time on machine m, respectively. Therefore, Constraints (18) and (19) guarantee that the completion time of job i on machine m must respect the estimated failure time and repair time $[t_m^f, t_m^R]$.

$$h\left(t_{i,k}^{End}\right) - Z_{i,k,m}\cdot\left(t_m^f + t_m^R\right) \geq D_{i,k} \qquad \forall i,k,m \tag{18}$$

$$h\left(t_{i,k}^{End}\right) - H\cdot Z_{i,k,m} \leq t_m^f \qquad \forall i,k,m \tag{19}$$

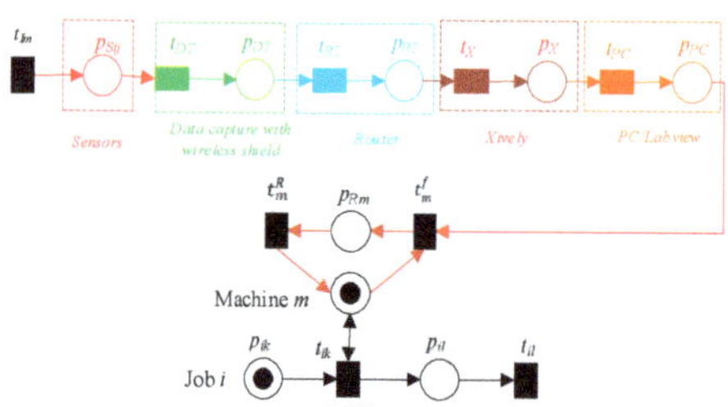

Figure 12. Representation of maintenance period for machine in an FMS based on IoT.

In addition, Constraints (20)–(22) guarantee the non-negativity of the variables.

$$h\left(t_{i,k}^{Start}\right) \geq 0 \qquad \forall i,k \tag{20}$$

$$h\left(t_{i,k}^{End}\right) \geq 0 \qquad \forall i,k \tag{21}$$

$$D_{i,k} \geq 0 \qquad \forall i,k \tag{22}$$

Finally, the objective function is to find the optimal firing sequence (denoted as δ) of transitions according to the time intervals related to the completion time of firing $h\left(t_{i,k}^{End}\right)$ of the last transition, which corresponds to operation k for job i in the last machine m. Thus, the objective function is defined as follows:

$$Min\left(\max_{\substack{i=1,\,2,\ldots,\Omega\\ k=1,\,2,\ldots,\Psi i}} h\left(t_{i,k}^{End}\right)\right) \tag{23}$$

3.3. Validation of the MILP Formulation

This section illustrates the MILP formulation's validity. In other words, it implies that viable solutions to the formulation above relate to feasible schedules in the sense that all jobs must be completed while meeting the precedence connection and conflict-free constraint on shared jobs and machines.

Theorem 1. *A viable schedule for the UFMS scheduling problem is constructed by a solution that satisfies all of the constraints from (7) to (25).*

Proof. In the MILP formulation, there are five main types of decision variables: $h\left(t_{i,k}^{Start}\right)$ and $h\left(t_{i,k}^{End}\right)$ are real variables, while $X_{i,k,m}$, $Y_{i,k,l}$, and $O_{i,k,m,j,l}$ are binary (integer) variables. Consider the ζ to be a viable solution. The ζ is guaranteed to meet the precedence relationship by Constraint (24). The value of $X_{i,k,m}$ in ζ indicates that operation k for job i has been assigned to machine m. The value of $Y_{i,k,l}$ shows that operation l is processed after operation k for job i. The value of $O_{i,k,m,j,l}$ specifies the order of precedence between operations k and l that use the shared machine m. Using the values of $X_{i,k,m}$, $Y_{i,k,l}$, and $O_{i,k,m,j,l}$, a scheduled timed PN can be constructed. □

Using the developed MILP formulation, the optimal solution for the UFMSs' scheduling problem is converted into a system controller to ensure that the resulting sequence is executed. The final TPN model is reconstructed based on the optimal firing sequences and the number of shared resources (machine tools) in the systems, as presented in Algorithm 2.

Algorithm 2: *Controlled TPN model based on the optimal firing sequences*

Input: *A TPN-based IoT (N_S, M_{S0});*
Output: *The optimal firing sequences δ and scheduling controllers for shared resources;*

1. *Create a MILP model based on the timed PN;*
2. *Solve the MILP model;*
3. *Compute the optimal firing sequences δ;*
4. *Generate copies of the shared places for an optimal firing sequence δ, i.e., the number of shared resource copies = the number of end transitions in δ – the number of the shared machines in the model;*
5. Generate the incidence matrix for the δ;
6. *Design scheduling controllers for shared resources based on the above matrix.*

Theorem 2. *The controlled TPN model based on the optimal firing sequences is live.*

Proof. It is necessary to demonstrate that all transitions in T in (N, M_o) are live. For all $t_{i,k} \in T$; if $\forall p_{i,k} \in {}^\bullet t_{i,k}$, $M(p_{i,k}) > 0$, then $t_{i,k}$ can fire. Therefore, the controlled net (N, M_o) is live. □

3.4. Illustrative Example

Consider the FMS example presented in Figure 13 to demonstrate the modeling of timed PNs. It consists of three machines (resources) (m_1, m_2, and m_3) that can process four kinds of parts (part 1, part 2, part 3, and part 4) in a sequential manner; two robots, r_1 and r_2 (material handling equipment); and a load/unload station. Table 3 displays the sequence of job-production operations, the duration of each operation, and the machine that can be used for each operation. As shown in Table 3, each part involves a set of operations. For instance, part 1 contains three operations that sequentially require m_1, m_2, and m_3. Thus, part 1 can be completed sequentially using machines m_1, m_2, and m_3 with corresponding processing times of 3, 5, and 6, respectively.

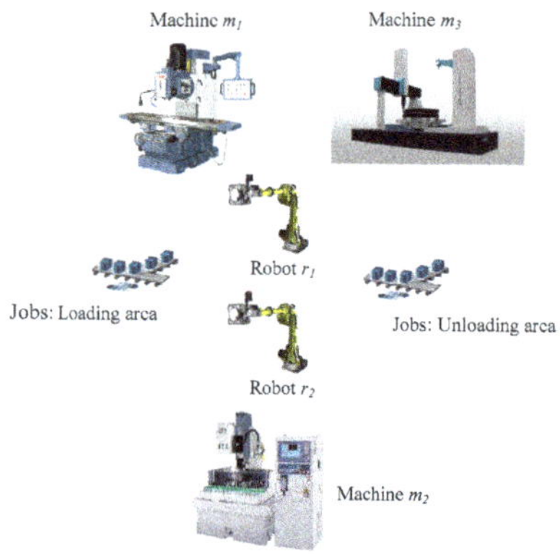

Figure 13. An illustrative example of an FMS.

Table 3. Scheduling data of an FMS shown in Figure 13.

Part	(Operation, Operation Time, Machine)		
1	(1, 3, M1)	(2, 5, M2)	(3, 6, M3)
2	(1, 8, M3)	(2, 4, M1)	(3, 9, M2)
3	(1, 2, M1)	(2, 7, M3)	(3, 5, M1)
4	(1, 3, M1)	(2, 2, M2)	(3, 6, M2)

Figure 14 illustrates the timed PN model for this system, and the transitions of the TPN model are given in Table 4. The parameters of machine maintenance are displayed in Table 5.

Table 4. Description of places and transitions of the TPN model shown in Figure 13.

Places	Description	Transitions	Description
p_{11}	Process 1 for part 1	t_{11}	End process 1 of part 1
p_{12}	Process 2 for part 1	t_{12}	End process 2 of part 1
p_{13}	Process 3 for part 1	t_{13}	End process 3 of part 1
p_{21}	Process 1 for part 2	t_{21}	End process 1 of part 2
p_{22}	Process 2 for part 2	t_{22}	End process 2 of part 2
p_{23}	Process 3 for part 2	t_{23}	End process 3 of part 2
p_{31}	Process 1 for part 3	t_{31}	End process 1 of part 3
p_{32}	Process 2 for part 3	t_{32}	End process 2 of part 3
t_{33}	Process 3 for part 3	t_{33}	End process 3 of part 3
p_{41}	Process 1 for part 4	t_{41}	End process 1 of part 4
p_{42}	Process 2 for part 4	t_{42}	End process 2 of part 4
p_{43}	Process 3 for part 4	t_{43}	End process 3 of part 4

Table 4. *Cont.*

Places	Description	Transitions	Description
p_{m1}	Machine 1 is available	t_{B1}^{PM}	Start PdM on machine 1
p_{m2}	Machine 2 is available	t_{B2}^{PM}	Start PdM on machine 2
p_{m3}	Machine 3 is available	t_{B3}^{PM}	Start PdM on machine 3
p_1^{PM}	PdM on machine 1	t_{E1}^{PM}	End PdM on machine 1
p_2^{PM}	PdM on machine 2	t_{E2}^{PM}	End PdM on machine 2
p_3^{PM}	PdM on machine 3	t_{E3}^{PM}	End PdM on machine 3

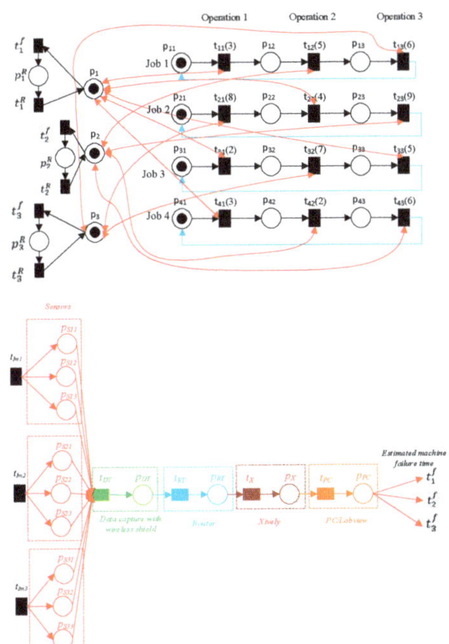

Figure 14. A timed PN model of the system shown in Figure 13.

Table 5. PdM parameters for machines presented in Figure 13.

Parameter	Machine 1	Machine 2	Machine 3
t_m^f (hr)		Based on sensor system	
t_m^R (hr)	3	3	3

A portion of a mill dataset is used to assess the tool's condition in a specific operational setting, as conducted by [57]. Data gathering involves the sampling of data using three different kinds of sensors: a "vibration sensor", an "acoustic emission sensor", and a "current sensor". Figure 14 illustrates that the system utilizes nine sensors to detect tool failure. These sensors include the current sensors p_{s11}, p_{s12}, and p_{s13}, which monitor changes in both the direct current (DC) and alternating current (AC) spindle motors for machines 1–3. Additionally, the system uses the accelerometer sensors p_{s21}, p_{s22}, and p_{s23}, which measure vibrations in the table and spindle for machines 1–3. Finally, the acoustic emission sensors p_{s31}, p_{s32}, and p_{s33} are used to measure how acoustic stress waves affect the table and spindle. This helps find a tool break for machines 1–3. The sensors initially collect the data and transmit it to the "Wi-Fi wireless shield", the "wireless router", Xively, and ultimately to the PC/lab view. The data collected by the sensors and transmitted to the Internet are illustrated in Figure 15. It allows for the analysis and visualization of the data to detect possible patterns or significant occurrences. When the sensor data surpass a certain threshold, the TPN model initiates the system's response.

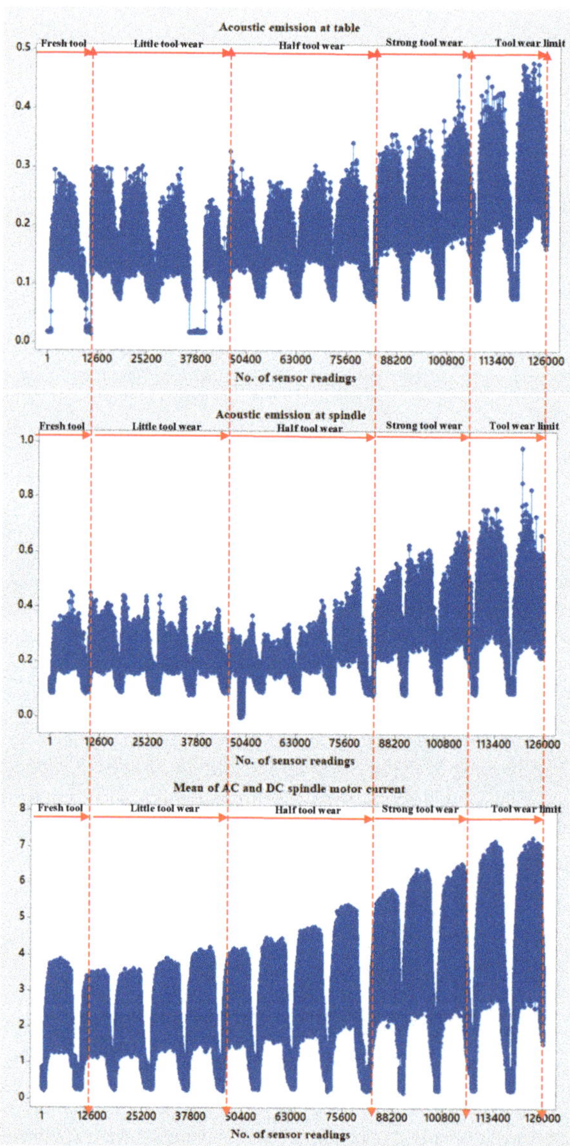

Figure 15. Data collected by sensors to monitor the condition of the tool used in the model depicted in Figure 14.

With this example, the Lingo solver solves the MILP model derived from Figure 14. Table 6 provides the optimal results. Figure 16 depicts Gantt charts for the results obtained by the proposed MILP model.

Table 6. Optimal results for the illustrative example.

Job	Operation	Time	Machine	$t_{i,k}^{Start}$	$t_{i,k}^{End}$
1	1	3	1	5	8
	2	5	2	13	18
	3	6	3	18	24
2	1	8	3	0	8
	2	4	1	8	12
	3	9	2	21	30
3	1	2	1	3	5
	2	7	3	8	15
	3	5	1	16	21
4	1	3	1	0	3
	2	2	2	3	5
	3	6	2	5	11

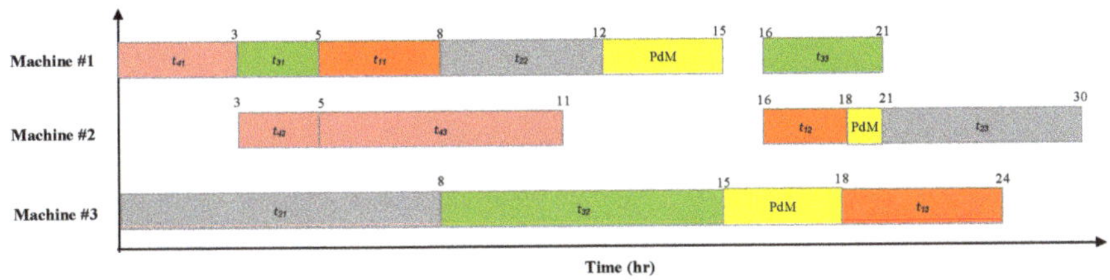

Figure 16. The optimal schedule for the illustrative example.

The optimal sequence of start firing transitions is: $t_{41}, t_{21}, t_{31}, t_{42}, t_{11}, t_{43}, t_{22}, t_{32}, t_{12}, t_{33}, t_{13}, t_{23}$. The controlled TPN model is designed and displayed in Figure 17 using Algorithm 2 and the optimal firing sequences achieved. Note that the arcs connecting the last place to the first place are excluded in each sequential system and in the estimation resource failure time networks, as depicted in Figure 17.

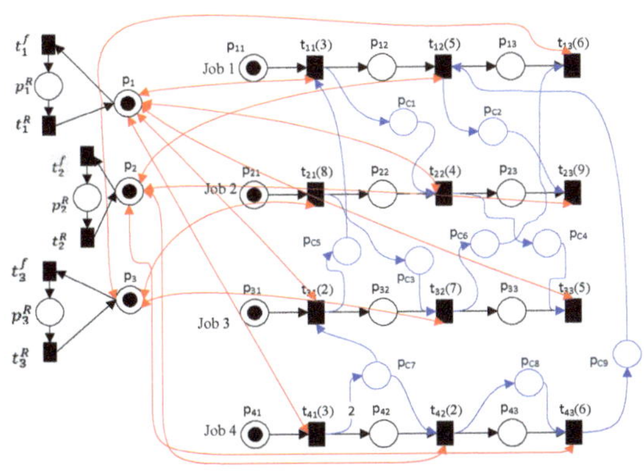

Figure 17. The controlled TPN for the model shown in Figure 14.

4. Computational Experiments

To validate the effectiveness of the proposed optimal scheduling and modeling approach, MILP generation software was developed. This software was created using the MATLAB-based GPenSIM tool [58]. The GPenSIM tool is particularly useful for modeling target-scheduling problems and exporting Excel documents. The Lingo solver utilizes the Petri net structure and tokens extracted from Excel documents as input data for the MILP process, which is subsequently used to solve the problem. The developed methodology is applied to systematically create MILP instances for the dynamic scheduling problem of unreliable flexible manufacturing systems. The developed methodology necessitates that users represent their desired problems using a Petri net. However, the modeling process is easy, provided the users possess expertise in the specific problem domain and are familiar with the fundamental concepts of Petri nets.

Case study 1: The system has two machines and three components. Parts 1, 2, and 3 were subjected to 3, 2, and 3 processes, respectively. The data related to this example are depicted in Tables 7 and 8.

Table 7. Scheduling data of case study 1.

Part	(Operation, Operation Time, Machine)		
1	(1, 3, M1)	(2, 5, M1)	(3, 6, M2)
2	(1, 8, M1)	(2, 4, M2)	-
3	(1, 9, M2)	(2, 2, M1)	(3, 7, M2)

Table 8. PdM parameters for the machines presented in case study 1.

Parameter	Machine	
	1	2
t_m^f (hr)	Based on sensor system	
t_m^R (hr)	3	3

The inputs for the derived mathematical model are acquired from the TPN model, as previously explained. The resultant model is solved using the Lingo solver to obtain the optimal schedule. The minimal compilation time is 34 h. Figure 18 depicts Gantt charts for the results obtained using the proposed MILP model. The optimal sequence of start firing transitions is: t_{11}, t_{31}, t_{21}, t_{32}, t_{22}, t_{12}, t_{32}, t_{13}, t_{12}. The controlled TPN model is displayed in Figure 19 using Algorithm 2 and the optimal firing sequences achieved.

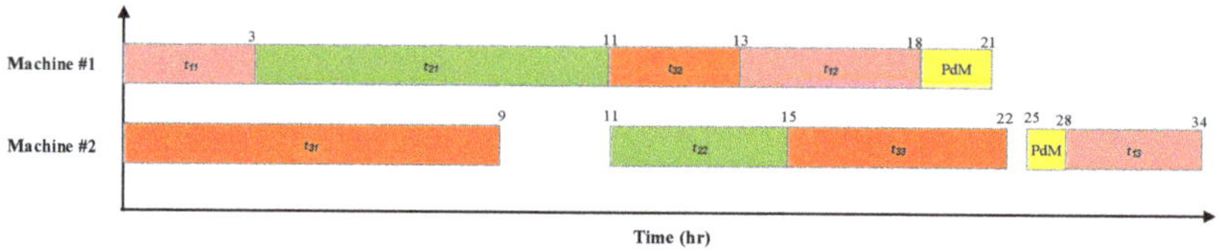

Figure 18. The optimal schedule for case study 1.

Case study 2: The system has four machines and five components. Each component is subjected to a series of four processes. The data related to this example are depicted in Tables 9 and 10.

The minimal compilation time is 25 h. Figure 20 depicts Gantt charts for the results obtained using the proposed MILP model. The optimal sequence of start firing transitions is: t_{11}, t_{51}, t_{41}, t_{21}, t_{12}, t_{42}, t_{31}, t_{43}, t_{52}, t_{22}, t_{13}, t_{53}, t_{23}, t_{32}, t_{44}, t_{14}, t_{33}, t_{54}, t_{24}, t_{34}. The controlled TPN model is displayed in Figure 21.

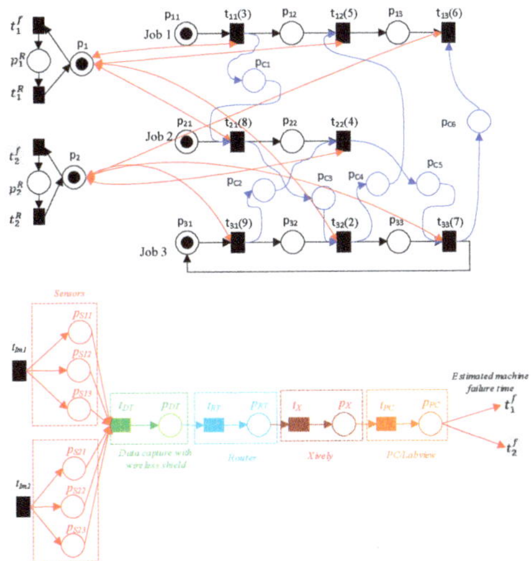

Figure 19. The controlled TPN for the system in case study 1.

Table 9. Scheduling data of case study 2.

Part	(Operation, Operation Time, Machine)			
1	(1, 2, M1)	(2, 2, M3)	(3, 2, M4)	(4, 3, M2)
2	(1, 2, M1)	(2, 2, M4)	(3, 3, M3)	(4, 4, M2)
3	(1, 5, M1)	(2, 3, M2)	(3, 3, M3)	(4, 2, M4)
4	(1, 3, M3)	(2, 2, M4)	(3, 6, M2)	(4, 3, M4)
5	(1, 3, M2)	(2, 4, M3)	(3, 5, M1)	(4, 6, M4)

Table 10. PdM parameters for the machines presented in case study 2.

Parameter	Machine			
	1	2	3	4
t_m^f (hr)	Based on sensor system			
t_m^R (hr)	2	3	2	3

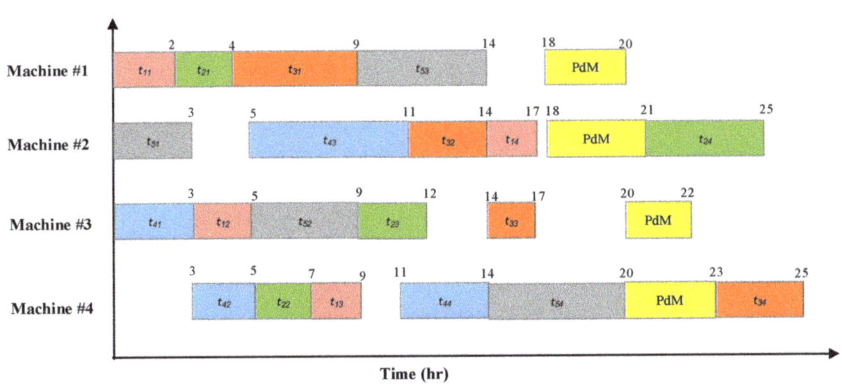

Figure 20. The optimal schedule for case study 2.

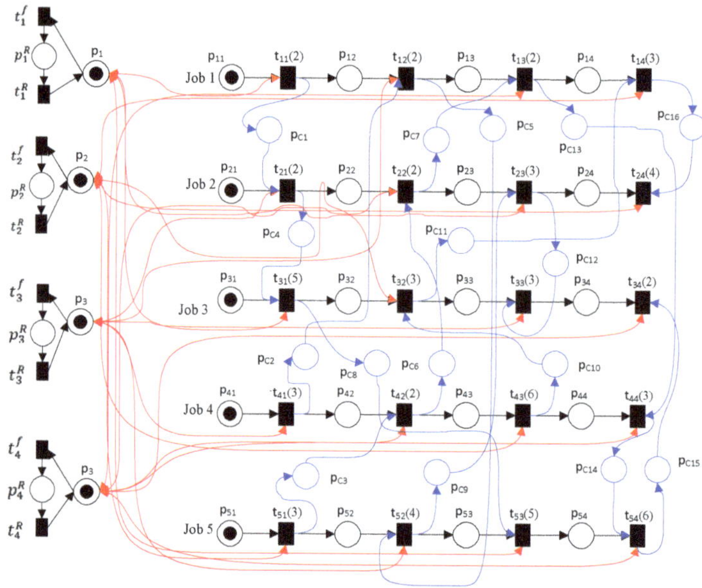

Figure 21. The controlled TPN for the system in case study 2.

The proposed approach is applicable to problems of large scale. The number of controllers will be augmented in proportion to the increase in shared resources, and the final result of the developed mathematical model might require the use of heuristic methods to minimize computational time. Despite the fact that most FMS scheduling problems are small, the proposed approach is applied to address larger problems. In order to validate the effectiveness of the proposed approach for more complex, dynamic scheduling problems with unreliable flexible manufacturing systems, it is implemented on a larger scale.

Case study 3: The system consists of 10 different types of parts and 6 machines. Each component is subjected to a series of five processes. The data related to this scenario are depicted in Tables 11 and 12. Figure 22 depicts the optimal schedule for this problem. The minimal cycle time is 64 units of time.

Table 11. Scheduling data of case study 3.

Part	(Operation, Operation Time, Machine)				
1	(1, 3, M1)	(2, 5, M2)	(3, 6, M3)	(4, 7, M4)	(5, 8, M5)
2	(1, 8, M3)	(2, 4, M1)	(3, 9, M2)	(4, 6, M6)	(5, 5, M6)
3	(1, 2, M1)	(2, 7, M3)	(3, 5, M1)	(4, 4, M5)	(5, 6, M4)
4	(1, 3, M1)	(2, 2, M2)	(3, 6, M2)	(4, 3, M4)	(5, 2, M5)
5	(1, 5, M6)	(2, 4, M3)	(3, 6, M1)	(4, 7, M2)	(5, 5, M5)
6	(1, 6, M5)	(2, 1, M6)	(3, 8, M6)	(4, 9, M4)	(5, 7, M5)
7	(1, 6, M1)	(2, 5, M2)	(3, 9, M3)	(4, 8, M4)	(5, 7, M6)
8	(1, 5, M5)	(2, 4, M4)	(3, 3, M3)	(4, 4, M2)	(5, 5, M1)
9	(1, 6, M6)	(2, 7, M6)	(3, 7, M4)	(4, 6, M1)	(5, 5, M5)
10	(1, 6, M4)	(2, 5, M3)	(3, 4, M2)	(4, 5, M5)	(5, 6, M6)

Table 13 illustrates the size and efficiency of the proposed approach across all of the studies. It demonstrates that the proposed approach includes a greater number of variables and constraints when the problem size is larger. Furthermore, the CPU processing time undergoes a significant increase as the problem size is increased. Thus, the scheduling problem is a complicated and NP-complete problem with a combinatorial structure.

Table 12. PdM parameters for the machines presented in case study 3.

Parameter	Machine 1	2	3	4	5	6
t_m^f (hr)			Based on sensor system			
t_m^R (hr)	5	3	1	1	2	2

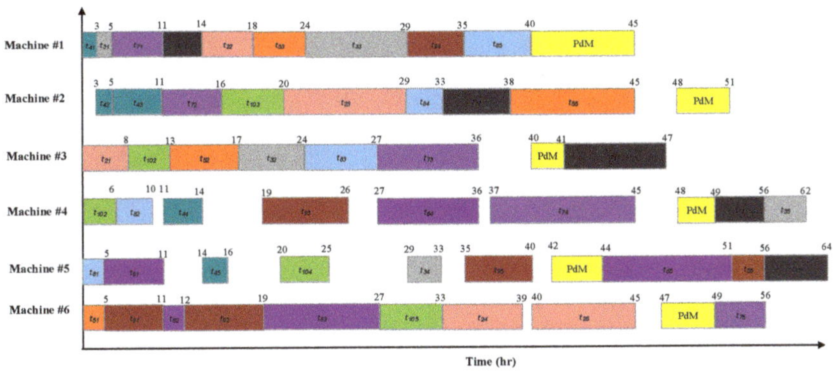

Figure 22. The optimal schedule for case study 3.

Table 13. Performance of the developed model in all case studies.

Parameter	Case Study 1	2	3
No. of Parts	3	5	10
No. of Machines	2	4	6
CPU time (second)	2	8	5160

5. Conclusions

This paper presents a method for generating MILP models using timed Petri nets. The method is designed to address the dynamic scheduling problems in UFMSs' complex industrial conditions. The objective is to construct TPN models using the IoT for the PdM configuration of mechanical equipment in the dynamic scheduling problem of UFMSs to determine the optimal production cycle time. The effectiveness of the proposed approach is demonstrated through several computational instances. The main contributions of the proposed method are as follows:

1. It can methodically build MILP instances based on TPN models for UFMSs under complex operational conditions such as concurrency, conflicts, resource sharing, and sequential processes, which can be applied in practical scenarios.
2. It provides a hybrid approach that integrates the TPN, the PdM configuration, and the IoT to predict and prevent early mechanical equipment failures in UFMSs, thereby avoiding disruptions to the scheduled operations of other UFMSs.
3. The optimal solution for the UFMS scheduling problem obtained from MILP can be converted into a system controller to guarantee the execution of the resulting sequence.

The following is a list of the significant results of the proposed method:

1. The computational results are better in terms of the quality of the obtained solutions.
2. It enables faster, optimal decision-making in DSP and solves instances for up to 10 parts and 6 machines.
3. It is capable of handling all unreliable machines in the UFMSs, and it is perfectly adequate to reach a solution in an acceptable CPU time.

While the current optimization solvers, such as LINGO and high-performance computers, are capable of solving large-scale scheduling problems with significant CPU time, there is a need for approximate approaches such as GA, ACO, PSO, SA, and TS to address these scheduling problems. This will be the focus of future research in this field. Furthermore, our objective is to expand our approach to address optimization issues that involve uncertainty. Stochastic TPNs can be applied to address uncertainty and formulate stochastic optimization problems based on PN models.

Author Contributions: Conceptualization, H.K. and A.A.-A.; methodology, H.K. and A.A.-A.; software, H.K. and A.A.-A.; validation, H.K., A.A.-A., K.N.A., A.D. and W.A.; resources, H.K. and A.A.-A.; data curation, H.K. and A.A.-A.; writing—original draft preparation, H.K. and A.A.-A.; writing—review and editing, H.K., A.A.-A., K.N.A., A.D. and W.A.; supervision, A.A.-A.; project administration, A.A.-A.; funding acquisition, A.A.-A. All authors have read and agreed to the published version of the manuscript.

Funding: The authors are grateful to the Raytheon Chair for Systems Engineering for funding.

Data Availability Statement: All the data are available in the manuscript.

Acknowledgments: The authors would like to thank the Raytheon Chair for Systems Engineering for funding.

Conflicts of Interest: The authors declare no conflict of interest.

References

1. Kaid, H.; Al-Ahmari, A.; Li, Z.; Ameen, W. An Improved Synthesis Method Based on ILPP and Colored Petri Net for Liveness Enforcing Controller of Flexible Manufacturing Systems. *IEEE Access* **2022**, *10*, 68570–68581. [CrossRef]
2. Chen, Y.; Li, Z.; Al-Ahmari, A. Nonpure Petri net supervisors for optimal deadlock control of flexible manufacturing systems. *IEEE Trans. Syst. Man Cybern. Syst.* **2012**, *43*, 252–265. [CrossRef]
3. Stecke, K.E. Design, planning, scheduling, and control problems of flexible manufacturing systems. *Ann. Oper. Res.* **1985**, *3*, 1–12. [CrossRef]
4. Lewis, H.R. Michael R. Garey and David S. Johnson. *Computers and intractability. A guide to the theory of NP-completeness*. W. H. Freeman and Company, San Francisco1979, x + 338 pp. *J. Symb. Log.* **1983**, *48*, 498–500. [CrossRef]
5. Modibbo, U.M.; Raghav, Y.S.; Hassan, M.; Mijinyawa, M. A Critical Review on the Applications of Optimization Techniques in the UN Sustainable Development Goals. In Proceedings of the 2021 2nd International Conference on Intelligent Engineering and Management (ICIEM), London, UK, 28–30 April 2021; pp. 572–576.
6. Modibbo, U.M.; Ali, I.; Ahmed, A. Multi-objective optimization modelling for analysing sustainable development goals of Nigeria: Agenda 2030. *Environ. Dev. Sustain.* **2021**, *23*, 9529–9563. [CrossRef]
7. Emrouznejad, A. *Big Data Optimization: Recent Developments and Challenges*; Springer: Berlin/Heidelberg, Germany, 2016; Volume 18.
8. Brouer, B.D.; Karsten, C.V.; Pisinger, D. Big data optimization in maritime logistics. In *Big Data Optimization: Recent Developments and Challenges*; Springer: Berlin/Heidelberg, Germany, 2016; pp. 319–344.
9. Abualigah, L.; Gandomi, A.H.; Elaziz, M.A.; Hamad, H.A.; Omari, M.; Alshinwan, M.; Khasawneh, A.M. Advances in meta-heuristic optimization algorithms in big data text clustering. *Electronics* **2021**, *10*, 101. [CrossRef]
10. Ibaraki, T. Integer programming formulation of combinatorial optimization problems. *Discret. Math.* **1976**, *16*, 39–52. [CrossRef]
11. Kantor, I.; Robineau, J.-L.; Bütün, H.; Marechal, F. A mixed-integer linear programming formulation for optimizing multi-scale material and energy integration. *Front. Energy Res.* **2020**, *8*, 49. [CrossRef]
12. Al-Ahmari, A.; Kaid, H.; Li, Z.; Wu, N.; El-Tamimi, A.-A.; Qiao, Y. A New MINLP Continuous Time Formulation for Scheduling Optimization of Oil Refinery with Unreliable CDUs. *Math. Probl. Eng.* **2022**, *2022*, 1298495. [CrossRef]
13. Liu, J.; MacCarthy, B.L. A global MILP model for FMS scheduling. *Eur. J. Oper. Res.* **1997**, *100*, 441–453. [CrossRef]
14. Sankar, S.S.; Ponnanbalam, S.; Rajendran, C. A multiobjective genetic algorithm for scheduling a flexible manufacturing system. *Int. J. Adv. Manuf. Technol.* **2003**, *22*, 229–236. [CrossRef]
15. T Taghavifard, M.; Heydar, M.; S Mousavi, S. A genetic algorithm for scheduling flexible manufacturing cells. *J. Appl. Sci.* **2009**, *9*, 97–104. [CrossRef]
16. Candan, G.; Yazgan, H.R. Genetic algorithm parameter optimisation using Taguchi method for a flexible manufacturing system scheduling problem. *Int. J. Prod. Res.* **2015**, *53*, 897–915. [CrossRef]
17. Sharma, D.; Singh, V.; Sharma, C. GA based scheduling of FMS using roulette wheel selection process. In Proceedings of the International Conference on Soft Computing for Problem Solving (SocProS 2011), Roorkee, India, 20–22 December 2011; Volume 2, pp. 931–940.
18. Udhayakumar, P.; Kumanan, S. Some metaheuristic approaches for optimising tardiness of job and tool in a flexible manufacturing system. *Int. J. Adv. Oper. Manag.* **2012**, *4*, 219–252. [CrossRef]

19. Udhayakumar, P.; Kumanan, S. Sequencing and scheduling of job and tool in a flexible manufacturing system using ant colony optimization algorithm. *Int. J. Adv. Manuf. Technol.* **2010**, *50*, 1075–1084. [CrossRef]
20. Jerald, J.; Asokan, P.; Prabaharan, G.; Saravanan, R. Scheduling optimisation of flexible manufacturing systems using particle swarm optimisation algorithm. *Int. J. Adv. Manuf. Technol.* **2005**, *25*, 964–971. [CrossRef]
21. Zhang, C.Y.; Li, P.; Rao, Y.; Guan, Z. A very fast TS/SA algorithm for the job shop scheduling problem. *Comput. Oper. Res.* **2008**, *35*, 282–294. [CrossRef]
22. Tiwari, M.K.; Kumar, S.; Shankar, R. Solving part-type selection and operation allocation problems in an FMS: An approach using constraints-based fast simulated annealing algorithm. *IEEE Trans. Syst. Man Cybern. Part A: Syst. Hum.* **2006**, *36*, 1170–1184. [CrossRef]
23. Pitts, R., Jr.; Ventura, J.A. Scheduling flexible manufacturing cells using Tabu Search. *Int. J. Prod. Res.* **2009**, *47*, 6907–6928. [CrossRef]
24. Sarma, U.; Kant, S.; Rai, R.; Tiwari, M. Modelling the machine loading problem of FMSs and its solution using a tabu-search-based heuristic. *Int. J. Comput. Integr. Manuf.* **2002**, *15*, 285–295. [CrossRef]
25. Gholami, M.; Zandieh, M. Integrating simulation and genetic algorithm to schedule a dynamic flexible job shop. *J. Intell. Manuf.* **2009**, *20*, 481–498. [CrossRef]
26. Pergher, I.; Frej, E.A.; Roselli, L.R.P.; de Almeida, A.T. Integrating simulation and FITradeoff method for scheduling rules selection in job-shop production systems. *Int. J. Prod. Econ.* **2020**, *227*, 107669. [CrossRef]
27. Thenarasu, M.; Rameshkumar, K.; Di Mascolo, M.; Anbuudayasankar, S.P. Multi-criteria scheduling of realistic flexible job shop: A novel approach for integrating simulation modelling and multi-criteria decision making. *Int. J. Prod. Res.* **2024**, *62*, 336–358. [CrossRef]
28. Bihari, M.; Kane, P. Evaluation and Improvement of Makespan Time of Flexible Job Shop Problem Using Various Dispatching Rules—A Case Study. In Proceedings of the Advances in Mechanical Engineering: Select Proceedings of ICAME 2020, Makassar, Indonesia, 14–15 October 2020; pp. 611–617.
29. Balog, M.; Dupláková, D.; Szilágyi, E.; Mindaš, M.; Knapcikova, L. Optimization of time structures in manufacturing management by using scheduling software Lekin. *TEM J.* **2016**, *5*, 319.
30. Murata, T. Petri nets: Properties, analysis and applications. *Proc. IEEE* **1989**, *77*, 541–580. [CrossRef]
31. Richard, P. Modelling integer linear programs with Petri nets. *RAIRO-Oper. Res.* **2000**, *34*, 305–312. [CrossRef]
32. Nakamura, M.; Tengan, T.; Yoshida, T. A Petri net approach to generate integer linear programming problems. *IEICE Trans. Fundam. Electron. Commun. Comput. Sci.* **2019**, *102*, 389–398. [CrossRef]
33. Tuncel, G.; Bayhan, G.M. Applications of Petri nets in production scheduling: A review. *Int. J. Adv. Manuf. Technol.* **2007**, *34*, 762–773. [CrossRef]
34. Wu, N.; Chu, F.; Chu, C.; Zhou, M. Petri net modeling and cycle-time analysis of dual-arm cluster tools with wafer revisiting. *IEEE Trans. Syst. Man Cybern. Syst.* **2012**, *43*, 196–207. [CrossRef]
35. Wu, N.; Zhou, M.; Chu, F.; Chu, C. A Petri-net-based scheduling strategy for dual-arm cluster tools with wafer revisiting. *IEEE Trans. Syst. Man Cybern. Syst.* **2013**, *43*, 1182–1194. [CrossRef]
36. Qiao, Y.; Wu, N.; Zhou, M. Scheduling of dual-arm cluster tools with wafer revisiting and residency time constraints. *IEEE Trans. Ind. Inform.* **2013**, *10*, 286–300. [CrossRef]
37. Qiao, Y.; Wu, N.; Zhou, M. A Petri net-based novel scheduling approach and its cycle time analysis for dual-arm cluster tools with wafer revisiting. *IEEE Trans. Semicond. Manuf.* **2012**, *26*, 100–110. [CrossRef]
38. Zhou, M.; McDermott, K.; Patel, P.A. Petri net synthesis and analysis of a flexible manufacturing system cell. *Syst. Man Cybern. IEEE Trans.* **1993**, *23*, 523–531. [CrossRef]
39. Zhou, M.; Chiu, H.-S.; Xiong, H.H. Petri net scheduling of FMS using branch and bound method. In Proceedings of the IECON'95—21st Annual Conference on IEEE Industrial Electronics, Orlando, FL, USA, 6–10 November 1995; pp. 211–216.
40. Artigues, C.; Roubellat, F. A Petri net model and a general method for on and off-line multi-resource shop floor scheduling with setup times. *Int. J. Prod. Econ.* **2001**, *74*, 63–75. [CrossRef]
41. Mejía, G.; Odrey, N.G. An approach using Petri nets and improved heuristic search for manufacturing system scheduling. *J. Manuf. Syst.* **2005**, *24*, 79–92. [CrossRef]
42. Zhang, W.; Freiheit, T.; Yang, H. Dynamic scheduling in flexible assembly system based on timed Petri nets model. *Robot. Comput.-Integr. Manuf.* **2005**, *21*, 550–558. [CrossRef]
43. Kim, Y.W.; Suzuki, T.; Narikiyo, T. FMS scheduling based on timed Petri Net model and reactive graph search. *Appl. Math. Model.* **2007**, *31*, 955–970. [CrossRef]
44. Lee, J.; Lee, J.S. Heuristic search for scheduling flexible manufacturing systems using lower bound reachability matrix. *Comput. Ind. Eng.* **2010**, *59*, 799–806. [CrossRef]
45. Wang, Q.; Wang, Z. Hybrid heuristic search based on Petri net for FMS scheduling. *Energy Procedia* **2012**, *17*, 506–512. [CrossRef]
46. Kammoun, M.A.; Ezzeddine, W.; Rezg, N.; Achour, Z. FMS scheduling under availability constraint with supervisor based on timed Petri nets. *Appl. Sci.* **2017**, *7*, 399. [CrossRef]
47. Xu, G.; Chen, Y. Petri-Net-Based Scheduling of Flexible Manufacturing Systems Using an Estimate Function. *Symmetry* **2022**, *14*, 1052. [CrossRef]

48. Ravidas, S.; Lekidis, A.; Paci, F.; Zannone, N. Access control in Internet-of-Things: A survey. *J. Netw. Comput. Appl.* **2019**, *144*, 79–101. [CrossRef]
49. Ray, P.P. A survey on Internet of Things architectures. *J. King Saud Univ. Comput. Inf. Sci.* **2018**, *30*, 291–319.
50. Asghari, P.; Rahmani, A.M.; Javadi, H.H.S. Internet of Things applications: A systematic review. *Comput. Netw.* **2019**, *148*, 241–261. [CrossRef]
51. Cachada, A.; Barbosa, J.; Leitão, P.; Alves, A.; Alves, L.; Teixeira, J.; Teixeira, C. Using internet of things technologies for an efficient data collection in maintenance 4.0. In Proceedings of the 2019 IEEE International Conference on Industrial Cyber Physical Systems (ICPS), Taipei, Taiwan, 6–9 May 2019; pp. 113–118.
52. Niyonambaza, I.; Zennaro, M.; Uwitonze, A. Predictive maintenance (Pdm) structure using internet of things (iot) for mechanical equipment used into hospitals in Rwanda. *Future Internet* **2020**, *12*, 224. [CrossRef]
53. Kaid, H.; Al-Ahmari, A.; Alqahtani, K.N. Fault Detection, Diagnostics, and Treatment in Automated Manufacturing Systems Using Internet of Things and Colored Petri Nets. *Machines* **2023**, *11*, 173. [CrossRef]
54. Mobley, R.K. *Maintenance Fundamentals*; Elsevier: Amsterdam, The Netherlands, 2011.
55. Iadanza, E.; Gonnelli, V.; Satta, F.; Gherardelli, M. Evidence-based medical equipment management: A convenient implementation. *Med. Biol. Eng. Comput.* **2019**, *57*, 2215–2230. [CrossRef]
56. Kaid, H.; Al-Ahmari, A.; Li, Z.; Ameen, W. Deadlock Control and Fault Detection and Treatment in Reconfigurable Manufacturing Systems Using Colored Resource-Oriented Petri Nets Based on Neural Network. *IEEE Access* **2021**, *9*, 84932–84947. [CrossRef]
57. Agogino, A.; Goebel, K.; Mill Data Set. BEST lab, UC Berkeley. NASA Ames Prognostics Data Repository. NASA Ames: Moffett Field, CA, USA, 2007. Available online: https://ti.arc.nasa.gov/project/prognostic-data-repository (accessed on 1 December 2022).
58. Davidrajuh, R. *Modeling Discrete-Event Systems with GPenSIM: An Introduction*; Springer: Berlin/Heidelberg, Germany, 2018.

machines

MDPI

Article

Enhancing Industrial Process Control: Integrating Intelligent Digital Twin Technology with Proportional-Integral-Derivative Regulators

Austeja Dapkute *, Vytautas Siozinys, Martynas Jonaitis, Mantas Kaminickas and Milvydas Siozinys

Company "Energy Advice", LT-44175 Kaunas, Lithuania; vytautas.siozinys@energyadvice.lt (V.S.); martynas.jonaitis@energyadvice.lt (M.J.); mantas.kaminickas@energyadvice.lt (M.K.); milvydas.siozinys@energyadvice.lt (M.S.)

* Correspondence: austeja@energyadvice.lt; Tel.: +370-602-30887

Abstract: This paper explores the integration of intelligent digital twin technology with PID regulators in industrial process control utilizing smart meter data. It presents a novel approach involving the creation of mathematical models to simulate real-time system behavior, thereby enhancing the PID control loop. The focus is on the development of specialized IT infrastructure to support this integration, which includes data acquisition, processing, and control optimization. This integration aims to not only improve control system efficiency but also introduce a robust predictive maintenance framework, offering significant benefits across a wide range of industrial applications.

Keywords: digital twin; intelligent digital twin; PID control optimization; state estimator; biomass boiler; process control

1. Introduction

In the relentless pursuit of efficiency, industrial systems are increasingly required to tackle challenges such as aging components, production variability, and the management of nonlinear dynamics. The paramount goal is to ensure the stability and adaptability of closed-loop systems, particularly in the face of resource scarcity. As mass-produced products inherently undergo aging and transformation, there is a pressing need for methodologies that dynamically tune controller parameters to accommodate characteristic fluctuations. Moreover, many industrial systems inherently display nonlinear behavior, complicating the control process.

In this context, Reinforcement Planned Control (RPC) emerges as a potent solution for nonlinear systems. It offers an intuitive approach, particularly when applied to proportional–integral–derivative (PID) controls, known for their low computational demand. Consequently, in the industrial landscape, there is a discernible inclination towards RPC-augmented PID controls, favored for their ease of implementation in mass-produced controllers. An integral objective of our research is to innovate in this domain by developing a method for designing RPC-enhanced PID controls directly.

However, it is crucial to acknowledge that traditional PID controls, despite being a cornerstone for process control, often falter in dynamically evolving environments. They may not adequately address unexpected disturbances caused by continuously changing operational contexts, which often necessitates manual intervention. This is particularly problematic in scenarios where immediate fluctuations—such as those induced by varying atmospheric temperatures—can adversely impact industrial processes. This gap underscores the significance of our research, aimed at enhancing industrial process control: integrating intelligent digital twin technology with PID regulators. By melding the precision of intelligent digital twin technology with the robustness of PID controls, we envision a transformative leap in industrial process management.

Additionally, our exploration delves into data-based control, leveraging input/output datasets to compute optimal parameters. Nonetheless, this approach is not devoid of challenges; a limited dataset can precipitate overfitting, thereby jeopardizing the stability of closed systems—a predicament reminiscent of the hurdles faced in machine learning, particularly in the development of nonlinear controllers for nonlinear systems. To navigate this complexity, we propose an adaptive, learning-based PID model, integrating the foresight of intelligent digital twin technology to refine its learning process. The implementation of this adaptive, learning PID, fortified with digital twin insights, constitutes a pivotal objective of our research, aiming to bring forth a new paradigm in industrial process control.

This article introduces EA-SAS, a newly developed intelligent digital twin platform, and presents a systematic exploration of the integration of intelligent digital twin technology with PID regulators in industrial process control, utilizing smart meter data. It begins with a literature analysis, situating our research within the broader academic context. Subsequent sections discuss the intelligent digital twin and PID technology integration IT infrastructure, including our developed Data Collector, Data Reader, Data Writer subsystems and PID loop integration approach. We present a high-level architecture overview, which addresses the complex demands of data processing within digital twins, a cornerstone for achieving operational efficiency and adaptability in contemporary industrial settings. Then, we describe the basics of the intelligent digital twin computational methodology for biomass boiler modeling and the state estimator technique.

These optimizations, facilitated by the integration of intelligent digital twin technology with the PID control loop, significantly enhance the system's adaptability and operational efficiency.

2. Related Work

2.1. Autonomous Manufacturing

In the literature review, a few real-world examples of autonomous management in manufacturing processes were identified. One of the cases involved an experimental solution deployed by Yokogawa Electric Corporation and JSR Corporation, detailed in the article from 22 March 2022, titled "In a World First, Yokogawa and JSR Use AI to Autonomously Control a Chemical Plant for 35 Consecutive Days" [1]. This study reported a successful field trial where artificial intelligence was utilized for the autonomous control of a chemical plant process, a claimed global first. In this instance, artificial intelligence (AI) autonomously managed distillation column valves at ENEOS Materials Corporation's chemical plant in Japan for 35 days.

In the article "AI used to control process manufacturing operations," [2] the creator of the control solution, PhD. Hiroaki Kanokogi, noted that AI management in industrial process plants is not as advanced as AI technology used in predictive maintenance. External atmospheric temperature changes can significantly impact many processes, necessitating complex control of temperature, pressure, and flow rate to avoid potential undesirable chemical or physical reactions. Typically, a proportional–integral–derivative (PID) control mathematical algorithm is used for process control. However, it may not adequately handle unexpected disturbances caused by continually changing environmental models, often requiring human intervention in situations where short-term changes due to varying atmospheric temperatures could negatively impact processes.

In this experiment, Yokogawa utilized artificial intelligence—the Factorial Kernel Dynamic Policy Programming (FKDPP) protocol, co-developed in 2018 by Yokogawa and Nara Institute of Science and Technology (NAIST). The experiment confirmed that reinforced learning AI could be safely applied in a real production environment and demonstrated that AI could independently manage complex processes previously controlled manually based on operator experience. During the field trial, the AI solution successfully dealt with the complex conditions required to maintain consistent product quality and the appropriate liquid level in the distillation column, maximizing the use of waste heat as a heat source. This led to stabilized quality, high product yield, and energy savings.

The article "Online Exclusive Technical Q&A: AI's use in chemical plant operations" [3] discussed the advantages of the next-generation Factorial Kernel Dynamic Policy Programming (FKDPP) control technology. In the industrial sector, a significant portion of AI technologies is made up of "problem analysis AI." This type of AI analyzes the presented data to detect predictive maintenance anomalies, forecast quality, or determine the cause of problems, typically supporting human decision-making. FKDPP represents a type of autonomous control AI, which seeks the optimal control model and then implements it. This uniqueness is highly beneficial, as this next-generation control technology can manage operations where existing methods (PID/APC) are insufficient, requiring manual control based on the qualifications of the plant personnel.

Recent advancements in autonomous manufacturing have extensively utilized digital twin technologies to enhance operational efficiencies and facilitate real-time decision-making. Specifically, studies have demonstrated various facets of digital twin applications: The authors of [4] presented methodologies for deriving project-specific digital twins tailored to industrial automation needs, focusing on modular integration and technology adaptability. An early implementation of digital twin technology was showcased in [5], where educational setups leveraged cloud computing and 5G networks for process automation. The authors of [6] analyzed digital twins for improving manufacturing systems' reconfiguration efficiencies, showing significant enhancements in setup time reductions. The authors of [7] explored deep reinforcement learning for optimizing manufacturing processes, contributing robust solutions to dynamic production challenges. Additionally, the authors of [8] discussed the implementation of flexible manufacturing cells under the ISO 23247 standard [9–12], emphasizing resilience and adaptability in production environments. The study by [13] also addressed safety and reliability in human–robot collaborations through advanced digital twin frameworks, which thus improved the predictive maintenance and interaction dynamics. Moreover, the authors of [14] focused on data-driven approaches to optimize control processes in manufacturing systems, integrating continuous feedback mechanisms for system enhancements. Lastly, the authors of [15] provided insights into AI-enhanced industrial automation systems using intelligent digital twins, showcasing the scalability and effectiveness of AI implementations in complex industrial settings.

2.2. Data-Driven Control

The article "Direct Data-Driven Control for Cascade Control System" [16] proposed a fresh perspective on cascade control systems, focusing on a direct synthesis method that eliminates the need for explicit plant modeling. The approach utilizes a direct data-driven PID design, extracting system dynamics directly from operational data, which is particularly advantageous in systems where mathematical models are difficult to obtain or are unreliable.

In "Designing Experiments for Data-Driven Control of Nonlinear Systems" [17], the authors delved into data-driven control solutions for complex systems. They highlighted the potential for such methods to adaptively learn and predict system behavior, thereby enabling more effective control strategies compared to traditional model-based approaches. The study demonstrated the viability of these methods through simulation and application in various control scenarios.

"Data-Driven Control in Autonomous Energy Systems" [18] focused on the theoretical aspects of data-driven control. The paper analyzed the stability and robustness of control systems that rely on data-driven techniques, providing a critical evaluation of their efficacy in real-world applications. The authors underscored the importance of robust data collection and processing methods to ensure the reliability of the control system.

These studies collectively underscore the shift towards leveraging real-time data for control purposes, reflecting a broader trend in automation and control engineering. Other papers in the field [19–26] corroborated these findings, showcasing data-driven control's adaptability to different system types and its capability to enhance system responsiveness and efficiency.

Multivariable data analysis, a cornerstone of statistical analysis, intricately explores the relationships and patterns among multiple variables. This foundational approach, detailed in "Multivariate Data Analysis" [27], is crucial for understanding complex data structures and making informed decisions. It employs a range of techniques, from exploratory factor analysis to structural equation modeling, offering insights into the interplay of variables and the underlying structure of data. These methods are invaluable across various domains, enabling a deeper comprehension and effective handling of multifaceted datasets.

2.3. Decision-Making Algorithms

By examining digital twin technology, our research emphasizes the utility of Markov Decision Processes (MDPs) and Dynamic Policy Programming (DPP) for enhancing decision-making in simulated environments. The literature highlights the importance of MDPs and DPP in complex decision-making frameworks [28,29], integrating these algorithms with digital twins to navigate intricate system dynamics effectively, thereby elevating process control and management in industrial settings.

MDPs facilitate the modeling of multi-state systems to determine optimal transitions, aiming for energy-efficient outcomes. This aligns with our goal to develop an intuitive platform for users to generate technological process states and associated rewards, like energy savings, showcasing MDP's role in energy optimization as a novel contribution.

Similarly, DPP tackles objectives akin to MDP but through the decomposition of problems into manageable sub-tasks, solved recursively. This method suits scenarios with overlapping tasks, enhancing problem-solving efficiency.

MDPs provide a solid structure for sequential decision-making under uncertainty, useful for various processes, from operations to AI. The adaptation of MDPs in Approximate Dynamic Programming (ADP) reflects their evolving frameworks for more stable, efficient learning processes in complex decision environments [30]. Policy iteration methods, promoting iterative policy refinement for optimal solutions, are gaining interest for their contribution to decision-making enhancement [31].

Garrett Thomas [32] explored MDPs' mathematical basis, crucial for discrete-time decision modeling, including state/action spaces and reward dynamics. The analysis stresses the discount factor's role in valuing future rewards and the policy's impact on decision-making, underlining policy optimization and convergence.

Rolf Iserman's contributions, particularly through his books "Fault-Diagnosis Systems" [33] and "Identification of Dynamic Systems" [34], provide foundational insights into process control and fault diagnosis integral to digital twin technology. Ref. [33] delved into fault detection and tolerance using statistical models and AI, which are crucial for enhancing system reliability—a key component of digital twins. Ref. [34], co-authored by Münchhof, focused on modeling dynamic systems, offering essential methodologies for accurate digital replication. While offering extensive theoretical and practical knowledge on system reliability and identification, Iserman's research could further benefit from addressing integration with newer technologies such as IoT and advanced predictive analytics, which are vital for modern digital twins' adaptability and efficiency.

To summarize, MDP focuses on current state-dependent future modeling, contrasting with DPP's broader application for scenarios where outcomes hinge on both current and historical states. MDPs excel in rapid decision scenarios like control theory and machine learning, whereas dynamic programming suits information-rich decision contexts, enabling more informed choices based on past influences.

2.4. PID Regulation and Auto-Tuning

For proportional–integral–derivative (PID) regulation and auto-tuning, the literature presents a multifaceted picture, showcasing various approaches and methodologies. The seminal work by Åström [35] outlines the ubiquity of PID controllers in industry, emphasizing their adaptability and the evolution of their tuning mechanisms due to advancements like microprocessors.

Furthermore, Liao and Zhao's research [36] presented an auto-tuning method for rotorcraft, integrating system identification procedures and a neural network to map controller parameters to performance, highlighting a novel approach to tuning cascade PID controllers.

"Multivariable Controller Tuning" by Johansson et al. [37] delved into the complexities of tuning individual loops in multivariable controllers. It explored the relation between a loop's performance and a specific row in the controller matrix, offering various interpretations and an algorithm for model estimation through relay feedback experiments. This process does not require prior system or controller knowledge, marking a significant stride in the tuning of interconnected control systems, especially where traditional models are not readily available or feasible to develop.

2.5. Digital Twin and Intelligent Digital Twin

With the advent of Industry 4.0, digital twin technology has become integral to enhancing both energy efficiency and process optimization. It necessitates comprehensive data analysis, including the mathematical modeling of physical systems and processes, forecasting, and applying statistical algorithms. Digital twin uses fresh monitoring data to represent the real-time state of the system and to estimate the future state [38]. For instance, the integration of digital twins across various industries has demonstrated significant benefits, including reduced operational costs, enhanced productivity, and superior decision-making capabilities [39]. Moreover, a review of Industry 4.0 from the perspective of automation and supervision systems identified digital twins as pivotal in the evolving architectures and functionalities of modern industrial systems, particularly in enhancing connectivity and intelligent automation [40].

An intelligent digital twin (IDT) is an extension of this definition, enhanced by the ability to observe its physical environment and to analyze and learn from it, so that existing models can be adapted or a real asset can interact with the resultant environment [15].

Traditional DTs do not anticipate future events or adjust actions to meet future goals. An IDT provides active assistance by working with and providing information as needed [41]. In our case here, we create an IDT that actively searches for the optimal conditions to achieve future goals and performs PID control corrections.

3. Intelligent Digital Twin Computational Methodology

Within this section, we describe the integration of combustion physics and state estimation algorithms to solve the equation system for biomass boiler operations. By applying decision-making algorithms, we define the optimal set points and PID coefficients to automate tuning and directly adjust the control system for enhanced regulation.

3.1. Biomass Boiler Modeling

The EA-SAS boiler intelligent digital twin uses continuous real-time data collection and cloud-based processing for mathematical modeling. This enables a real-time assessment of boiler efficiency, calculated using the ratio of produced heat ($Q_{produced}$) and the heat generated from fuel consumption (Q_{fuel}):

$$\eta_k = \frac{Q_{produced} + Q_{economizer}}{Q_{fuel(actual)}}. \quad (1)$$

The fuel calorific value is calculated as follows:

$$Q_{fuel(actual)} = Q_{fuel(LHV)} \cdot m_{efficient} - 0.02443 \cdot M, \quad (2)$$

Here, $Q_{fuel(LHV)}$ is the net calorific value (MJ/kg); $m_{efficient}$ is the efficient burnable mass; 0.02443 is the correction factor of the enthalpy of vaporization (constant pressure) for water (moisture) at 25 °C (MJ/kg per 1 w% of moisture); and M is the moisture content (w%).

The efficiently combusted biomass fuel mass is calculated as follows:

$$m_{\text{efficient}} = m_{\text{burnable}} \cdot \left(1 - \frac{A}{100}\right) = \left(1 - \frac{M}{100}\right) \cdot \left(1 - \frac{M}{100}\right), \tag{3}$$

Here, A is the ash content (w%).

The net calorific value can be calculated in two ways. If fuel composition is known, empirical equations may be used. The Mendeleev empirical equation used for determining the lower value is as follows:

$$Q_{\text{fuel(LHV)}} = 339C + 1035H - 109(O - S) - 25M, \tag{4}$$

Here, C is the carbon content (w%); H is the hydrogen content (w%); O is the oxygen content (w%); and S is the sulfur content (w%).

As it is problematic to determine the fuel composition, a lower heat value may be calculated:

$$Q_{\text{fuel(LHV)}} = qV, gr, d - 212.2 \cdot H - 0.8(O - N), \tag{5}$$

Here, qV, gr, d is a gross calorific value of fuel as determined by the biomass fuel sample.

The specific heat of biomass changes during devolatilization, and the particle temperature increases. The value of the raw biomass specific heat varies not only with temperature but also with water content [42].

The specific heat capacity of chemical species in the volatiles' non-condensable fraction is well known, with extensive information available in the literature. The most common method for approximating the relationship between specific heat and temperature is the NASA-type polynomial function, with up to seven coefficients, which are usually specified for two temperature ranges (below and above 1000 K) [42], although the used method depends on the needed accuracy. Lanzafame and Mesina [43] proposed the following logarithmic polynomial equation for specific heat at constant pressure:

$$\tilde{c}_p(T^*) = a_0 + a_1 \ln(T^*) + a_2[\ln(T^*)]^2 + \ldots + a_5[\ln(T^*)]^5, \tag{6}$$

Here, T^* is the flue gas temperature and a_k is the polynomials provided in Table 1 of [43].

Table 1. EA-SAS Data Hub VM computational requirements.

Operating System	Window Server	Linux
Version	-	Ubuntu Server 22 LTS
CPU	Dual core processor with virtualization	Dual core processor with virtualization
Storage *	70 GB SSD	50 GB SSD
Memory	6 GB	4 GB

* Additional space may be needed according to the amount of collected information.

The enthalpy of the mixture is calculated as follows:

$$h_{\text{mix}}(T^*) = \frac{\sum_{k=1}^{n.gases} x_k(T^*)\tilde{h}_k(T^*)}{\sum_{k=1}^{n.gases} x_k(T^*)M_k}, \tag{7}$$

Here, M is molecular weight, x is the molar fraction, and $\tilde{h}_k$ is specific enthalpy.

Then, the isobaric specific heat of a mixture can be calculated as follows:

$$c_{p,\text{mix}}(T^*) = \frac{\sum_{k=1}^{n.gases} x_k \tilde{c}_{p,k}(T^*) + \sum_{k=1}^{n.gases} \frac{dx_k}{dT^*}\tilde{h}_k(T^*)}{\sum_{k=1}^{n.gases} x_k(T^*)M_k}, \tag{8}$$

The amount of heat produced in the economizer by the water content change from d to d_{H_2O} can be calculated as follows:

$$Q_{economizer} = C_{P_{flue\ gas}} \cdot T_{flue\ gas} + \left(L_{td} + C_{pH_2O} \cdot T_{flue\ gas}\right) d - C_{P_{flue\ gas\_eco}} - \left(L_{td\_eco} + C_{pH_2O} \cdot T_{flue\ gas\_eco}\right) d_{H_2O}, \quad (9)$$

Here, $T_{flue\ gas}$ is the temperature of the flue from the furnace; $T_{flue\ gas\_eco}$ is the temperature of the flue gas exhausted from the economizer; $C_{P_{flue\ gas}}$ and $C_{P_{flue\ gas\_eco}}$ are specific volumetric heat capacities in accordance with the $T_{flue\ gas}$ and $T_{flue\ gas\_eco}$ temperatures; and L_{td} and L_{td_eco} are the latent heat of evaporation.

The boiler heat transfer coefficient U is calculated using real-time boiler data according to Equation (10):

$$q = U \cdot A \cdot \Delta T_{LM}, \quad (10)$$

Here, q is the heat transfer rate (W); U is the overall heat transfer coefficient ($W/(m^2 \cdot K)$); A is the heat transfer surface area (m^2); and ΔT_{LM} is the logarithmic mean temperature difference.

3.2. State Estimator

In our research, we deploy an advanced computational framework for determining the energy balance within complex engineering systems, modeled by our digital twin technology. EA-SAS Cloud uses the special state estimator technique, which allows for the set of Functions (1)–(10) to be solved together with the metered value vector z:

$$z = h(x) + \text{error}, \quad (11)$$

where z represents the vector of measurements, x denotes the true state vector (set of unknown variables), $h(\cdot)$ signifies the nonlinear vector function relating the measurements to the states, and e is the measurement error vector with zero mean and a variance denoted by R_z.

To estimate the state vector x, we solve a minimization problem of the following form:

$$J(x) = \frac{1}{2}(z - h(x))' R_z^{-1} (z - h(x)), \quad (12)$$

Here, the weighted least-squares problem is framed within an iterative process that leverages the gradient of J(x), denoted by g(x), and a gain matrix G(x), which varies depending on whether we employ the Gauss–Newton or Newton–Raphson methods for minimization.

In employing the Gauss–Newton method, we linearize the nonlinear function h(x) through a Taylor expansion, leading to the following:

$$h = (x + \Delta x) \backslash R_z^{-1} H(x))^{-1} H(x)' R_z^{-1} \Delta z, \quad (13)$$

Here, $\Delta z = z - h(x)$ and $H(x)$ represents the Jacobian matrix of $h(x)$.

Alternatively, the Newton–Raphson method introduces second-order derivatives into the equation, adjusting the state vector correction as follows:

$$\Delta x = \left(H(x)' R_z^{-1} H(x) - \sum_{j=1}^{m} \gamma_j \Delta z_j \frac{\delta^2 h}{\delta x^2} \right)^{-1} H(x)' R_z^{-1} \Delta z \quad (14)$$

The correction to the state vector (Δx) is critical, ensuring that our digital twin's estimation aligns closely with the real-world energy flows at each node within the system. It is noteworthy to mention that the inclusion of second-order derivatives tends to have a negligible effect on the convergence of our state estimation models, as the term involving these derivatives can often be omitted without significant loss of accuracy.

Employing combustion equations and state estimators, our digital twin precisely calculates the system's energy equilibrium. This method improves the model accuracy and optimizes the operations via exact energy flow measurements. Subsequently, we discuss the specialized IT infrastructure designed to facilitate PID auto-tuning.

4. Intelligent Digital Twin PID Control IT Infrastructure

4.1. Data Hub

Data Hub is a program specially designed for data collection and writing to a client's control system or directly into equipment. Data Hub consists of Data Collector and Data Writer. Data Collector is an application responsible for collecting measurement data from configured Data Readers and for transferring data readings to the main EA-SAS Cloud server. Data Collector can be installed and configured on the premises of the customer IT infrastructure (when most of the data are collected via the local server) or in the Cloud (if the data can be accessed from an external network). The Data Collector configuration allows IT admin to change the frequency of the data collection from configured Data Readers. Data Collector serves as a data pusher via the REST API to periodically transfer data to the EA-SAS Cloud server. The monitoring server is then dedicated to monitoring the VM server resources.

Data Reader is a service that is the part of Data Collector. It is responsible for collecting data from the source (measurement device or server). Data Reader configurations (data update frequency, etc.) are performed in the Data Collector interface.

Data writing into the control system is executed via the Data Writer program. Data Writer is also a part of the EA-SAS Data Hub. The Data Writer program was designed for data writing from the main server to the required equipment (usually the control system). Data Writer configurations (setting the data writing frequency, etc.) are performed in the Data Collector environment. Data Writing sequences are depicted in Figure 1.

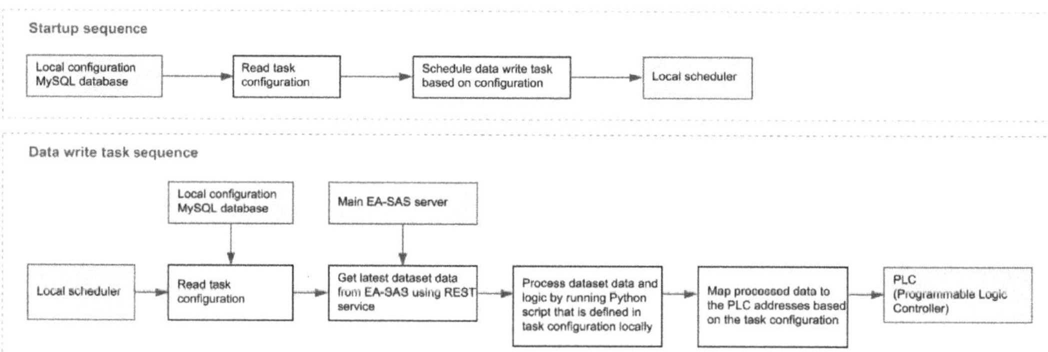

Figure 1. Example of a data writing sequence.

The startup sequence initiates the configuration by reading task parameters and preparing the scheduler for operation, while the data write task sequence involves regularly fetching, processing, and mapping data to the Programmable Logic Controllers (PLCs) for the execution of control tasks. The startup sequence (above) can be described as follows:

- Local Configuration MySQL (SQL-Structured Query Language) Database: The startup sequence commences with reading the task configuration from the local configuration MySQL database.
- Local Scheduler: A local scheduler then schedules the data write task based on the configuration it has read.

Data write task sequence (below):

- Local Configuration MySQL Database: Parallel to the startup sequence, during the data write task sequence, the system again reads the task configuration from the local MySQL database. The choice of MySQL, an open-source database, aligns with the principles of Industry 4.0, which emphasizes transparency, interoperability, and the strategic use of open-source technologies.
- Local Scheduler: The local scheduler operates as part of this sequence as well, although the specific action it takes in this phase is not detailed in the diagram.
- Main EA-SAS Server: The main EA-SAS server retrieves the latest dataset via a REST service.
- Python Script Execution: With the latest data acquired, a Python script is executed to process the dataset and apply the necessary logic. This script, which is defined within the local task configuration, is written in Python due to its open-source nature and its status as a standard programming language widely adopted across various industries. The choice of Python not only ensures flexible and robust data handling but also aligns with industry best practices in software development, enhancing the system's interoperability and adaptability.
- Mapping to PLC Addresses: In this example, processed data are mapped to PLC addresses following the guidelines specified in the task configuration. It is essential to recognize that the communication options within automation systems extend beyond PLCs, offering broad possibilities for integration. Available protocols include BACnet, HTTP, M-Bus, Modbus, OPC Data Access, MQTT, SQL, S7, and Wonderware, each providing unique features and capabilities for versatile system integration. Most of the PLCs have external communication capabilities. The implementation depends on the manufacturer, model, and series of the PLC and sometimes requires additional communication modules or protocol converters installed. The communication protocol also depends on the manufacturer, but newer models tend to support open-source industrial protocols and can be directly connected to the internet network.
- PLC: Finally, in the PLC, the mapped data are utilized as per the control logic requirements.

4.2. Computational Requirements for the EA-SAS Data Hub VM

The local server must have access to the same network as the object (boiler, cooling, drying, or other) control system and or any other measurement devices that can be connected. There must also be a constant internet connection to non-local IP addresses.

Usually, a virtual machine is created, but dedicated hardware is also acceptable. This server will be used for EA-SAS Data Hub installation. This software manages to read data from the control system and to then send it to the main EA-SAS Cloud. The requirements are presented at Table 1.

There are no particular requirements for VM IP address assignment. It can be static or dynamic. VM must have network access to the control system equipment that the data will be collected from.

The VM access requirements are listed here:

- EA-SAS Cloud, TCP, used for sending read data;
- EA-VPN, UDP, used for managing VM and accessing internal resources.
- EA-Docker register, TCP, used for installing/updating the EA-SAS Data Hub.
- APT registry. Official repositories are used for installing the packages required for the software;
- Additional network firewall requirements (which network ports will be used) depend on the protocol that EA-SAS Data Hub will use to read data from the control system. The protocol depends on the used control system implementation.

4.3. Intelligent Digital Twin PID Loop Integration

Referencing Figure 2, the system architecture integrates a digital twin within the PID control loop to refine the control strategy for a given process. The digital twin receives input data streams, which include measured data, set points, and additional system information.

This data assimilation enables the digital twin to create an accurate representation of the process, taking into account current conditions and desired outcomes.

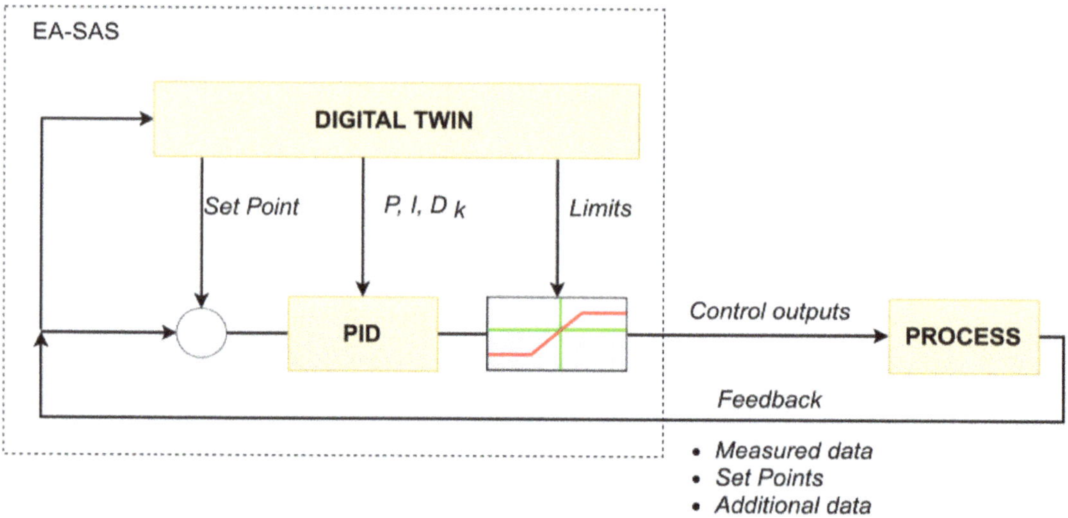

Figure 2. Intelligent digital twin PID loop integration.

Upon receiving the data, the digital twin performs real-time analyses and computes the necessary adjustments to the control set points. These computations are facilitated by the digital twin's capabilities to handle multiple parameters, represented by P_k, I_k, and D_k gains within the PID algorithm. Adjustments are made based on the process feedback and limits, ensuring that the PID controller's outputs are continuously tuned for system equilibrium.

The PID controller, equipped with the refined set points and limits provided by the digital twin, generates control outputs that drive the actual process towards the desired state. This closed-loop system is further fine-tuned through feedback loops, where the process outputs are continuously monitored and fed back into the digital twin, fostering an adaptive control environment.

The efficacy of this integrated approach lies in its ability to synthesize a high volume of data and translate it into actionable control strategies. By doing so, the PID loop, augmented with the digital twin's computational intelligence, offers a robust and responsive control mechanism, ensuring process stability and efficiency. This fusion of technologies underpins the contemporary drive towards smarter, data-driven process management within industrial systems.

4.4. High-Level Architecture Overview

The International Standardization Organization issued a series of standards (ISO 23247 [9–12],) that suggest a generic framework for digital twins for manufacturing. The EA-SAS intelligent digital twin architecture aligns with contemporary standards in digital twin technology, echoing the guidelines of ISO/IEC AWI Standard 30172 [44] and ISO/IEC AWI Standard 30173 [45]. The framework consists of the user entity for hosting software systems and interfaces, the digital twin entity for the digital representation and synchronization of observable manufacturing elements (OMEs), and the device communication entity for data interaction and device control. Each entity is further divided into sub-entities and functional entities, such as the data collection sub-entity for data acquisition and pre-processing and the device control sub-entity for actuation and operational control [38].

Within the high-level system architecture, depicted in Figure 3, the Energy Advice virtual machine (VM) hosts a specialized Data Hub, which comprises distinct Data Collector and Data Writer programs. These programs are fundamental to the execution of our methodological framework, executing two primary functions: the collection of process data and writing control data.

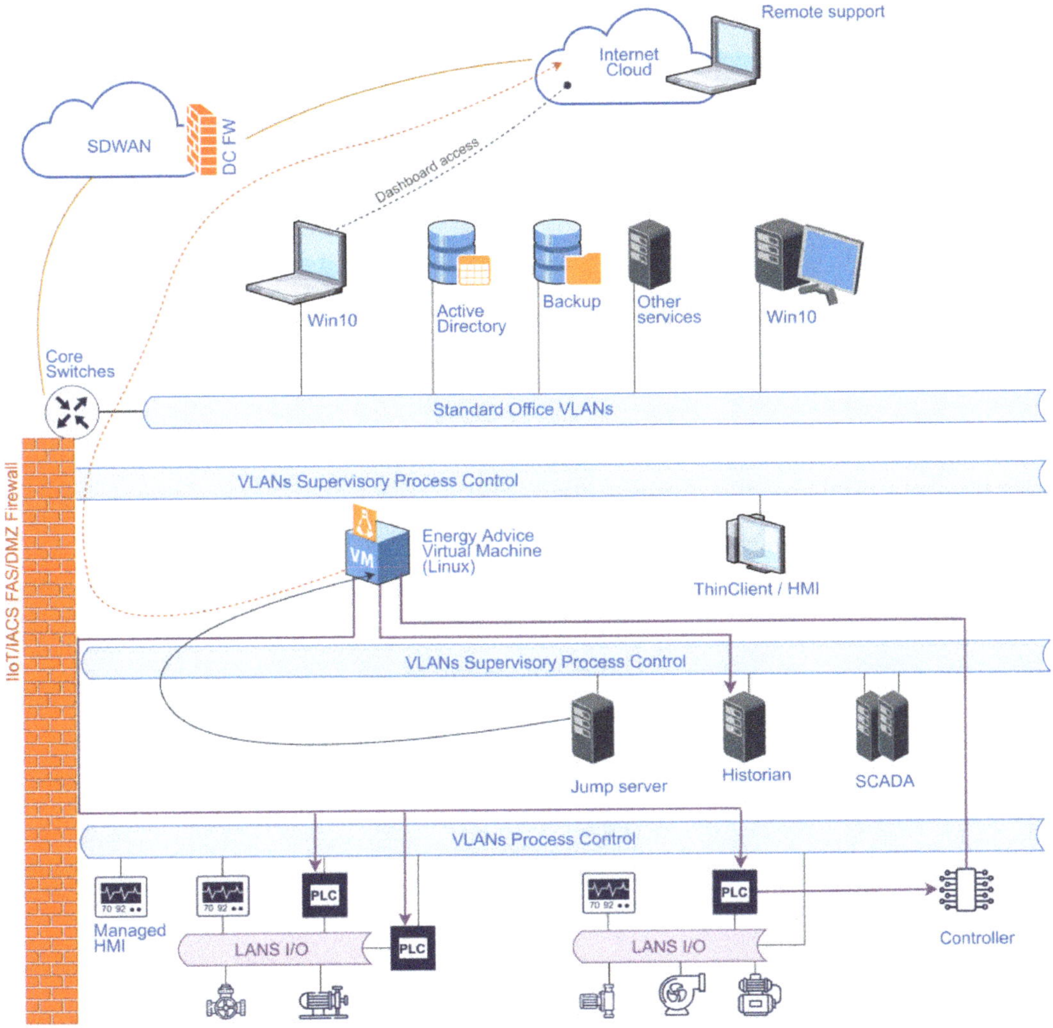

Figure 3. High-level architecture scheme.

The Data Collector program is tasked with interfacing with the process control network, utilizing various protocols to gather real-time operational data. These data are collected in a read-only mode, precluding any direct control interactions that could potentially affect process integrity. The VM operates within a VLAN specifically allocated for supervisory process control, which is segregated from the standard office VLANs to mitigate any cross-traffic interference, enhancing data security and system reliability.

Concurrently, the Data Writer program within the Data Hub processes the collected data, employing the established PID loop control methodology to ascertain the appropriate control set points. These set points are computed based on the analysis of live process data, reflecting the current state of the system. Once determined, the set points are communicated to the PLCs. The PLCs, in turn, implement these set points to adjust operational parameters in alignment with the desired control objectives.

Overall, the VM's operations within the control system are methodically structured to support the PID loop control approach, adhering to the scientific principles of data collection, process control theory, and network design, as detailed throughout our article. The VM's function is critical in synthesizing the diverse streams of process data into coherent control actions, embodying the core of our data-driven control methodology.

5. System Performance and Efficiency Insights

In this study, we demonstrated the successful deployment and operational efficacy of our digital twin of the biomass boiler within a dual-boiler system, comprising boilers K3 and K2, with only K3 currently operational. Illustrated in Figure 4, the application results validate our theoretical approach described in Section 3.1. 'Biomass Boiler Modeling,' revealing the system's operational dynamics. Through the EA-SAS platform, we achieved the autonomous adjustment of critical parameters like hot water temperature control for district heating, optimizing the fan efficiencies and fuel combustion rates. These results underscore the practicality and effectiveness of integrating digital twin technology with PID control loops, reflecting significant advancements in system control and operational precision.

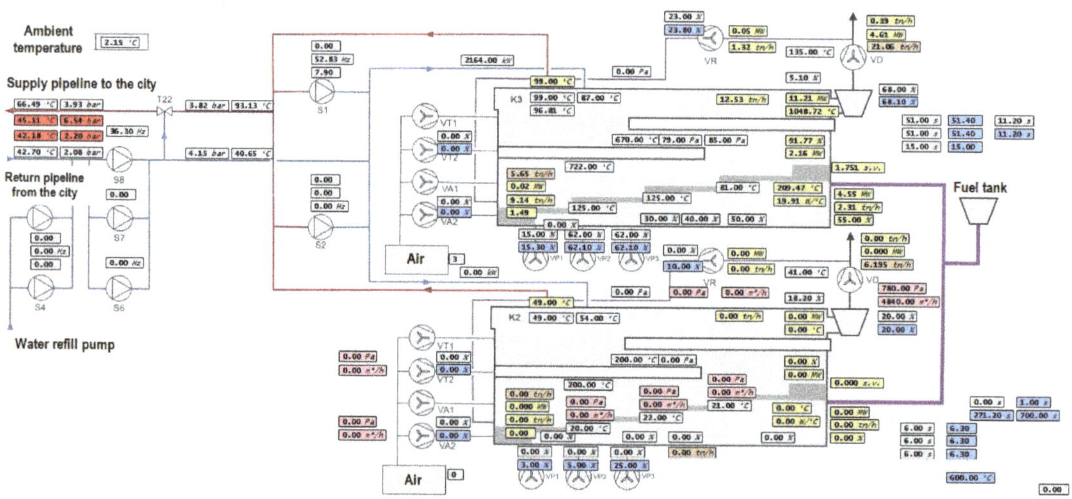

Figure 4. Visual representation of the calculation results on the EA-SAS intelligent digital twin system interface system.

The parameters automatically controlled without human intervention are listed in Table 2. All listed parameters are calculated in real time by the intelligent digital twin, considering the calculated calorific value of the fuel burned and the forecasted demand for thermal power. The calculated control commands are sent directly to the control system.

Undoubtedly, all parameters are closely related, and the created model of the boiler house and heat network is complex and comprehensive. It evaluates the influence of different components on the overall boiler house operation.

One of the main parameters that we aimed to stabilize in the biomass boiler operation is boiler output temperature. Stabilizing the biomass boiler water output temperature is critical for maintaining the efficiency of heat production. Figure 5 Illustrates the impact of control using the intelligent digital twin: post-implementation, temperature stability is significantly improved, reducing energy losses.

Table 2. Comparative analysis of ETL scheduling tools.

Equipment	EA-SAS Automatically Controlled Parameters	Description
Three-way valve	Temperature set point for the three-way valve, °C	Optimized temperature of the hot water supplied to the city heat network to minimize heat losses in the network while maintaining technological requirements.
Primary air fans Secondary air fans Tertiary air fans	Fans' efficiency, %	Optimizing the performance of primary, secondary, and tertiary air fans to ensure efficient fuel combustion.
Flue gas draft fans		Optimizing fan performance for optimal draft and combustion processes at the current boiler operation mode.
Flue gas recirculation fans		Optimizing fan performance to stabilize temperatures.
Fuel feeders	Feeder pause, s	Optimizing fuel layer and grate speed for efficient fuel combustion.
Grate	I grate speed, sII grate speed, sIII grate speed, s	

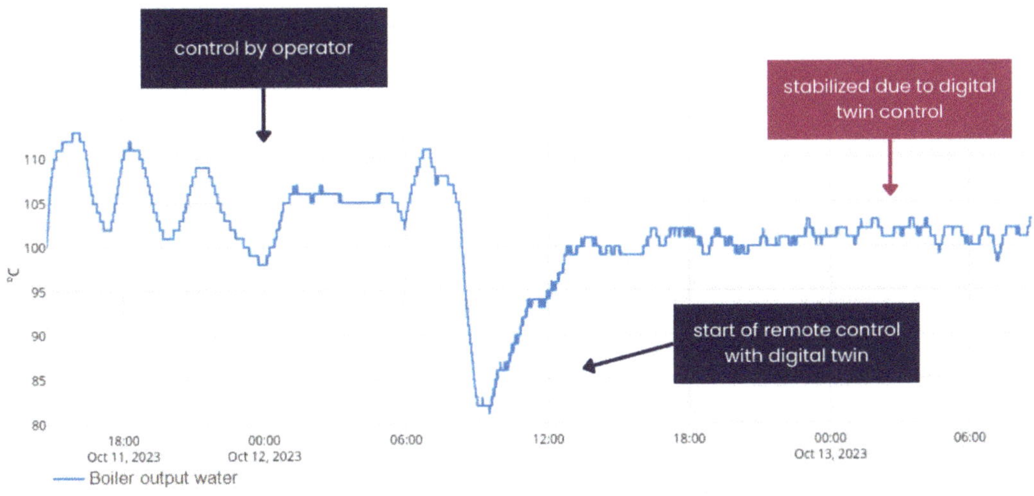

Figure 5. Boiler output water temperature stabilization.

By facilitating optimal remote control and proactive operational set point adjustments, the PID with intelligent digital twin in loop technology ensures precise temperature regulation. This results in increased energy efficiency, decreased fuel consumption, lower emissions, and improved longevity and reliability of the biomass power plant.

6. Conclusions

Our investigation into the integration of digital twin technology with PID control loops for biomass boiler operations demonstrates both methodological deployment and operational viability. This practical application, notably within a dual-boiler system where only K3 was in operation, highlights the real-world applicability and effectiveness of our approach. Recognizing the potential limitations such as data dependency and the continuous need for algorithmic refinement, we propose several avenues for future research.

First, further enhancements to the digital twin model should focus on increasing its predictive capabilities. This could involve incorporating more sophisticated machine learning algorithms to predict system behaviors and failures before they occur, thus improving the system's operational efficiency and reducing downtime.

Second, addressing the adaptability of the digital twin and PID control loops to manage and optimize operations in real-time across these varied settings will be essential. This could involve developing adaptive algorithms that can learn from the system's performance over time and adjust control strategies dynamically.

These research directions will not only push the boundaries of control technologies but also significantly contribute to the field of industrial process optimization, supporting a wider adoption and technological evolution in various sectors.

Author Contributions: Conceptualization, V.S. and M.S.; methodology, V.S., M.K. and M.J.; software, M.J.; validation, V.S., M.J. and A.D.; formal analysis, V.S., M.K. and M.J.; investigation, A.D. and M.J.; resources, M.J.; data curation, M.J.; writing—original draft preparation, A.D.; writing—review and editing, A.D.; visualization, A.D.; supervision, V.S.; project administration, A.D.; funding acquisition, V.S. All authors have read and agreed to the published version of the manuscript.

Funding: This research was funded by the Norwegian financial mechanism and the state budget of the Republic of Lithuania funds (grant number LT-07-1-EIM-K01-006).

Data Availability Statement: The data that support the findings of this study are not publicly available due to confidentiality agreements with our clients. These agreements prohibit the sharing of the data outside of the specific permissions granted for the research and publication of this manuscript.

Conflicts of Interest: All authors are employed by the Company "Energy Advice". The authors declare that the research was conducted in the absence of any commercial or financial relationships that could be construed as a potential conflict of interest.

References

1. Yokogawa & JSR Use AI to Autonomously Control Chemical Plant for 35 Days. Available online: https://www.automation.com/en-us/articles/march-2022/yokogawa-jsr-ai-autonomously-control-chemical (accessed on 12 January 2024).
2. AI Used to Control Process Manufacturing Operations | Control Engineering. Available online: https://www.controleng.com/articles/ai-used-to-control-process-manufacturing-operations/ (accessed on 12 January 2024).
3. Online Exclusive Technical Q&A: AI's Use in Chemical Plant Operations. Available online: https://www.hydrocarbonprocessing.com/news/2022/07/online-exclusive-technical-qa-ai-s-use-in-chemical-plant-operations (accessed on 12 January 2024).
4. Dittler, D.; Braun, D.; Müller, T.; Stegmaier, V.; Jazdi, N.; Weyrich, M. A Procedure for the Derivation of Project-Specific Intelligent Digital Twin Implementations in Industrial Automation. Available online: https://www.researchgate.net/publication/361743435 (accessed on 5 November 2023).
5. Kortela, J.; Nasiri, B.; Smirnov, A.; Lahnalammi, A.; Jämsä-Jounela, S.-L. Educational Setup for Service Oriented Process Automation with 5G Testbed. *IFAC-PapersOnLine* **2017**, *50*, 127–132. [CrossRef]
6. Talkhestani, B.A.; Weyrich, M. Digital Twin of manufacturing systems: A case study on increasing the efficiency of reconfiguration. *at-Automatisierungstechnik* **2020**, *68*, 435–444. [CrossRef]
7. Khdoudi, A.; Masrour, T.; El Hassani, I.; El Mazgualdi, C. A Deep-Reinforcement-Learning-Based Digital Twin for Manufacturing Process Optimization. *Systems* **2024**, *12*, 38. [CrossRef]
8. Wallner, B.; Zwölfer, B.; Trautner, T.; Bleicher, F. Digital Twin Development and Operation of a Flexible Manufacturing Cell using ISO 23247. *Procedia CIRP* **2023**, *120*, 1149–1154. [CrossRef]
9. *ISO 23247-1:2021*; Automation Systems and Integration—Digital Twin Framework for Manufacturing—Part 1: Overview and General Principles. International Organization for Standardization: Geneva, Switzerland, 2021. Available online: https://www.iso.org/standard/75066.html (accessed on 18 January 2024).
10. *ISO 23247-2:2021*; Automation Systems and Integration—Digital Twin Framework for Manufacturing—Part 2: Reference Architecture. International Organization for Standardization: Geneva, Switzerland, 2021. Available online: https://www.iso.org/standard/78743.html (accessed on 18 January 2024).
11. *ISO 23247-3:2021*; Automation Systems and Integration—Digital Twin Framework for Manufacturing—Part 3: Digital Representation of Manufacturing Elements. International Organization for Standardization: Geneva, Switzerland, 2021. Available online: https://www.iso.org/standard/78744.html (accessed on 18 January 2024).
12. *ISO 23247-4:2021*; Automation Systems and Integration—Digital Twin Framework for Manufacturing—Part 4: Information Exchange. International Organization for Standardization: Geneva, Switzerland, 2021. Available online: https://www.iso.org/standard/78745.html (accessed on 18 January 2024).

13. Wang, S.; Zhang, J.; Wang, P.; Law, J.; Calinescu, R.; Mihaylova, L. A deep learning-enhanced Digital Twin framework for improving safety and reliability in human–robot collaborative manufacturing. *Robot. Comput. Manuf.* **2024**, *85*, 102608. [CrossRef]
14. He, R.; Chen, G.; Dong, C.; Sun, S.; Shen, X. Data-driven digital twin technology for optimized control in process systems. *ISA Trans.* **2019**, *95*, 221–234. [CrossRef] [PubMed]
15. Jazdi, N.; Talkhestani, B.A.; Maschler, B.; Weyrich, M. Realization of AI-enhanced industrial automation systems using intelligent Digital Twins. *Procedia CIRP* **2020**, *97*, 396–400. [CrossRef]
16. Jianwang, H.; Ramirez-Mendoza, R.A.; Xiaojun, T. Direct Data-Driven Control for Cascade Control System. *Math. Probl. Eng.* **2021**, *2021*, 1465493. [CrossRef]
17. De Persis, C.; Tesi, P. Designing Experiments for Data-Driven Control of Nonlinear Systems. *IFAC-PapersOnLine* **2021**, *54*, 285–290. [CrossRef]
18. Örfler, F.D.; Ürich, Z. Data-Driven Control in Autonomous Energy Systems. Available online: http://people.ee.ethz.ch/~floriand/docs/Slides/Dorfler_GT_2020.pdf (accessed on 12 January 2024).
19. De Persis, C.; Tesi, P. Formulas for Data-Driven Control: Stabilization, Optimality, and Robustness. *IEEE Trans. Autom. Control.* **2019**, *65*, 909–924. [CrossRef]
20. Martin, T.; Schön, T.B.; Allgöwer, F. Guarantees for data-driven control of nonlinear systems using semidefinite programming: A survey. *Annu. Rev. Control* **2023**, *56*, 100911. [CrossRef]
21. Hou, Z.; Jin, S. A novel data-driven control approach for a class of discrete-time nonlinear systems. *IEEE Trans. Control Syst. Technol.* **2011**, *19*, 1549–1558. [CrossRef]
22. Control System Synthesis-Data-Driven Control Introduction to Data-Driven Control Predictive and Learning DDC Use of Local Models Use of Repetitive Experiments Robust DDC Using Convex Optimization Using Non-Convex Optimization Model Reference DDC. Available online: https://www.control.lth.se/fileadmin/control/Education/DoctorateProgram/ControlSystemsSynthesis/2020/Control_System_Synthesis___data_driven_control.pdf (accessed on 12 January 2024).
23. Selvi, D.; Piga, D.; Battistelli, G.; Bemporad, A. Optimal direct data-driven control with stability guarantees. *Eur. J. Control* **2021**, *59*, 175–187. [CrossRef]
24. MSafonov, G.; Holmes, S.; Doyle, A.C. Data-Driven Robust Control Design: Unfalsified Control. Available online: http://www.rto.nato.int/abstracts.asp (accessed on 12 January 2024).
25. Baggio, G.; Bassett, D.S.; Pasqualetti, F. Data-driven control of complex networks. *Nat. Commun.* **2021**, *12*, 1429. [CrossRef] [PubMed]
26. Tanaskovic, M.; Fagiano, L.; Novara, C.; Morari, M. Data-driven control of nonlinear systems: An on-line direct approach. *Automatica* **2017**, *75*, 1–10. [CrossRef]
27. Hair, J.F., Jr.; Black, W.C.; Babin, B.J.; Anderson, R.E. Multivariate Data Analysis, 7th ed. Available online: https://www.drnishikantjha.com/papersCollection/Multivariate%20Data%20Analysis.pdf (accessed on 12 January 2024).
28. Stamatakis, G.; Tragos, E.Z.; Traganitis, A. Energy Efficient Policies for Data Transmission in Disruption Tolerant Heterogeneous IoT Networks. *Sensors* **2018**, *18*, 2891. [CrossRef] [PubMed]
29. Tahir, I.; Nasir, A.; Algethami, A. Optimal Control Policy for Energy Management of a Commercial Bank. *Energies* **2022**, *15*, 2112. [CrossRef]
30. Bertsekas, D.P. Regular policies in abstract dynamic programming. *SIAM J. Optim.* **2017**, *27*, 1694–1727. [CrossRef]
31. Azar, M.G.; Gómez, V.; BKAPPEN, H.J.K. Dynamic Policy Programming. *J. Mach. Learn. Res.* **2012**, *13*, 3207–3245.
32. Thomas, G. Markov Decision Processes. 2020. Available online: https://ai.stanford.edu/ (accessed on 12 January 2024).
33. Isermann, R. *Fault-Diagnosis Systems: An Introduction from Fault Detection to Fault Tolerance*; Springer: Berlin/Heidelberg, Germany, 2006. [CrossRef]
34. Isermann, R.; Münchhof, M. *Identification of Dynamic Systems: An Introduction with Applications*; Springer: Berlin/Heidelberg, Germany, 2011. [CrossRef]
35. Åström, K.J. Control System Design. 2002. Available online: https://www.cds.caltech.edu/~murray/courses/cds101/fa02/caltech/astrom.html (accessed on 12 January 2024).
36. Liao, G.; Zhao, J. Auto-tuning for cascade PID height position controller of rotorcraft. *MATEC Web Conf.* **2019**, *277*, 01008. [CrossRef]
37. Johansson, K.H.; James, B.; Bryant, G.F.; Åström, K.J. Multivariable Controller Tuning. Available online: https://ieeexplore.ieee.org/document/703258 (accessed on 12 January 2024).
38. Dapkute, A.; Siozinys, V.; Jonaitis, M.; Kaminickas, M.; Siozinys, M. Digital Twin Data Management: Framework and Performance Metrics of Cloud-Based ETL System. *Machines* **2024**, *12*, 130. [CrossRef]
39. Singh, M.; Srivastava, R.; Fuenmayor, E.; Kuts, V.; Qiao, Y.; Murray, N.; Devine, D. Applications of Digital Twin across Industries: A Review. *Appl. Sci.* **2022**, *12*, 5727. [CrossRef]
40. Folgado, F.J.; Calderón, D.; González, I.; Calderón, A.J. Review of Industry 4.0 from the Perspective of Automation and Supervision Systems: Definitions, Architectures and Recent Trends. *Electronics* **2024**, *13*, 782. [CrossRef]
41. Grieves, M. Intelligent digital twins and the development and management of complex systems. *Digit. Twin* **2022**, *2*, 8. [CrossRef]
42. Popescu, F.; Mahu, R.; Ion, I.V.; Rusu, E. A Mathematical Model of Biomass Combustion Physical and Chemical Processes. *Energies* **2020**, *13*, 6232. [CrossRef]

43. Rosario, L.; Michele, M. Thermodynamic Property Models for Unburned Mixtures and Combustion Gases | Request PDF. *Int. J. Thermodyn.* **2006**, *9*, 73–80. Available online: https://www.researchgate.net/publication/42539893_Thermodynamic_Property_Models_for_Unburned_Mixtures_and_Combustion_Gases (accessed on 12 January 2024).
44. *ISO/IEC TR 30172:2023*; Internet of Things (loT)—Digital Twin—Use Cases. ISO: Geneva, Switzerland, 2023. Available online: https://www.iso.org/standard/81578.html (accessed on 12 January 2024).
45. *ISO/IEC 30173:2023*; Digital Twin—Concepts and Terminology. ISO: Geneva, Switzerland, 2023. Available online: https://www.iso.org/standard/81442.html (accessed on 12 January 2024).

MDPI AG
Grosspeteranlage 5
4052 Basel
Switzerland
Tel.: +41 61 683 77 34

Machines Editorial Office
E-mail: machines@mdpi.com
www.mdpi.com/journal/machines

www.ingramcontent.com/pod-product-compliance
Lightning Source LLC
Chambersburg PA
CBHW062059220926
51850CB00010B/2823
* 9 7 8 3 7 2 5 8 6 6 8 0 9 *